The Salvation of the Embryonic Person

Beginning with the Incarnation and the Immaculate Conception

Holiness comes to meet the Sacredness of Life!

Francis Etheredge

Foreword by Dr. Andrew Beards

Endorsed by Prof. Rev. Alberto Carrara

En Route Books and Media, LLC
Saint Louis, MO

ENROUTE
Make the time

En Route Books and Media, LLC
5705 Rhodes Avenue
St. Louis, MO 63109

Contact us at
contactus@enroutebooksandmedia.com

Cover Credit: Sebastian Mahfood using DALL-E

ISBN-13: 979-8-88870-612-1
Library of Congress Control Number:
Available online at https://catalog.loc.gov

TABLE OF CONTENTS

DR. ANDREW BEARDS: BIOGRAPHY

Dr Andrew Beards: B.A., M.A., Ph.L.,
Ph.D. (Calgary), Ph.D. (Lateran)

Andrew Beards is a lecturer in philosophy and tutor at Allen Hall Seminary, London, teaching on the Pontifical Degree offered by St Mary's University, Twickenham. He is the author of *Objectivity and Historical Understanding*, (Brookfield. VA & Aldershot: Ashgate, 1997), *Method in Metaphysics:* (Toronto: University of Toronto Press, 2008), *Insight and Analysis*, (New York and London: Continuum, 2010), *Philosophy the Quest for Truth and Meaning*, (Collegeville, Minn: Liturgical Press, 2010), *Lonergan, Meaning and Method: Philosophical Essays*, (New York and London: Bloomsbury Academic, 2016) and *The Challenge of Lonergan's Thought: The Reach of Critical Realism*, (New York and London: Bloomsbury Academic, 2025).

FOREWORD

In his address to the international diplomatic corps of the 9th January 2026, our new Holy Father, Pope Leo XIV, delivered a powerful message to the community of nations. In it, he underlined the true foundations of a just society in the family and the authentic social foundation of marriage: the indissoluble union between a man and a woman, open to life. He forcefully condemned abortion and euthanasia (assisted suicide) and other assaults upon the dignity of the human person, such as in vitro fertilization and the violation of consciences, even in democratic countries, as witnessed in attempts to force those in the medical profession to participate in such violations of human dignity. All these disastrous social ideologies are imposed, His Holiness maintained, by the Orwellian cultivation of 'woke language' which in the name of 'inclusivity' in fact excludes and marginalises those who would defend an authentic and objective natural law without which society is adrift in a sea of competing relativisms.

In the light of the Holy Father's clear teaching, we can be most grateful for this new book from the seasoned Catholic writer Francis Etheredge. The work, in fact, unpacks further much of what Pope Leo was saying in his recent address and defends it, drawing upon a wide range of literature from the Church's magisterium, particularly from Vatican II onwards. But his purview also includes very informative information on these and related ethical issues from publicly available surveys and data. The effect of such evidence is to bring home to us the horror of abortion, that silent holocaust ever in our midst. But this darkness is only examined in the true light of our salvation in Christ. Thus, the thematic counterpoint to this evil is the book's focus on the Incarnation of the Son of God and the Immaculate Conception of Our Lady – the proleptic beginnings of our salvation. In these mysteries of faith, we find the true dignity of human personhood from conception to natural death.

Among the themes of the book are the Divine mercy, which Etheredge rightly emphasises and points out has been particularly stressed in aspects of the recent magisterium. It is not that those committed to pro-life are vengeful Pharisees who wish to see women who have had abortions punished and vilified for their crimes. Rather, data, the witness of those who have experienced abortion and other sources of evidence, point to the fact that however truly difficult and morally complex the situation may be of those who feel driven to have an abortion – and we must also be honest in saying that for a good number such complex difficulties do not in reality obtain – not only is abortion never the solution insofar as it takes the life of the innocent, but that it is truly damaging to both the mother and father involved. Perhaps we need in our day not only to stress the Divine mercy, but also to reflect anew upon Divine justice; for mercy has no meaning without justice. Nor is it the case that Divine justice is in reality a concealed form of cruel vengeance, which we as enlightened twenty-first-century people have now exposed as such and have thus moved beyond. The Divine mercy and the Divine justice are brought together in the Divine wisdom.

The book serves to bring home to us that while each one of the areas of purported 'moral progress' in our society - abortion, contraception, a radical feminist critique of the nuclear family, the advocation of 'gay rights', the transgender agenda, assisted euthanasia, invitro fertilisation - is deeply problematic in itself, each and every case must also be seen as part of the whole, the whole of what St Pope John Paul II named 'the culture of death'. It is indeed a diabolical attack on an authentic anthropology of the human being, the human being in community. The richness of a Catholic anthropology of gift, the gift of each life and of communal human life as in the image of Trinitarian love, is the antidote Etheredge forcefully and persuasively argues, to this malaise. And indeed, there is growing and hopeful evidence of a younger generation in the West beginning to see this 'darkness made visible' and thus turning to the light, the light always present in the mystical body of Christ – even despite the tragic failures of so many of its members.

I believe readers will find this book replete with valuable theological insights that arise from the author's close engagement with Sacred Scripture, Tradition, and the magisterium of the Church. This is not to say that one may agree with all the points made in the work. I have reservations concerning the notion that Christ's redemption is already applied in the supernatural state of our first parents before the fall, and I am not yet wholly convinced of the ethics of abandoned embryo adoption. However, the stimulating way these ideas arise from the text should further encourage the reader to engage with it.

Dr. Andrew Beards,
Allen Hall Seminary,
London

PROF. REV. ALBERTO CARRARA

Prof. Rev. Alberto Carrara, LC, PhD
Dean of the Faculty of Philosophy
Pontifical *Athenaeum Regina Apostolorum* (Rome)
Member of the Pontifical Academy for Life:

An Endorsement:

A NECESSARY BOOK IN A TIME OF ANTHROPOLOGICAL BLINDNESS

The Salvation of the Embryonic Person by Francis Etheredge addresses one of the most dramatic and unresolved moral questions of our time: the fate of embryonic human beings generated through *in-vitro* fertilization and left in a condition of indefinite cryopreservation. This is no longer a marginal bioethical issue. It is a structural injustice produced by modern technoscientific culture and sustained by a deep anthropological confusion regarding the meaning of human life, embodiment, and generation. We live in a paradoxical age. On the one hand, human rights are constantly invoked as universal and inviolable. On the other hand, millions of human beings at the embryonic stage are treated as disposable biological material, suspended between existence and non-existence, deprived of any concrete future. Cryopreserved embryos represent a silent humanity, unseen, unheard, and structurally abandoned. Etheredge's book does not approach this reality primarily through technical, legal, or pragmatic categories. Instead, it proposes a radical shift in perspective:

the question of embryonic human life must be re-read in the light of the mystery of the Incarnation, the Annunciation, and the Immaculate Conception. This theological horizon is not an extrinsic or devotional addition to bioethics; it is a foundational anthropological key. It reminds us that Christianity begins not with an abstract moral norm but with God Himself's concrete assumption of human life. This is the book's first major strength: it restores the personal and ontological status of the human embryo, not as a "problem to be managed," but as a subject to be recognized.

The Human Embryo Beyond Biological Reductionism

A central issue in contemporary bioethical debates is the widespread reduction of the human embryo to a "cluster of cells," a "biological potential," or a pre-personal entity whose moral value depends on future development or functional capacities. This reductionism, biological, neurological, or functionalist, rests on a flawed anthropology in which dignity is no longer grounded in being, but in performance. Etheredge decisively challenges this view by recovering an ontological and personalist anthropology. From the very first moment of its existence, the human embryo is a human being with intrinsic dignity. This claim is not merely theological; it is coherent with a sound philosophical anthropology and with the constant teaching of the Catholic Church, from *Donum Vitae* to *Dignitas Personae*. The theological reflection on the Incarnation plays a decisive role here. If the Son of God assumed human nature from the first instant of His human conception, then no stage of human life can be considered pre-personal or morally negligible. The

Incarnation reveals that human life is worthy of communion with God from its very beginning. In this sense, Etheredge's work is not only a defense of embryonic life but also a critique of the broader cultural paradigm that instrumentalizes the human body and fragments personal identity. The embryo is not "something" that will later become "someone." The embryo is already someone.

The Original Injustice and the Absence of Ideal Solutions

One of the most intellectually honest aspects of this book is its acknowledgment of the original injustice that underlies the existence of cryopreserved embryos. The production of human embryos through IVF, followed by their selection, freezing, or abandonment, constitutes a moral disorder at its root. No subsequent action can fully repair this original wrong. As *Dignitas Personae* soberly states, the situation of abandoned frozen embryos constitutes a "situation of injustice which in fact cannot be resolved." This statement must be taken seriously. No ideal or morally "clean" solutions are available after the fact. Etheredge does not deny this tragic reality, nor does he attempt to dissolve it through rhetorical optimism. Instead, he poses the most difficult and unavoidable question: what moral responsibility do we have today toward these concrete human beings who already exist and who have been radically abandoned? Silence, inaction, or moral paralysis cannot be morally neutral responses. To do nothing is itself a choice, one that leaves these human beings without any possibility of fulfilling their natural destiny.

Adoption for Birth: An Extraordinary Remedy in an Extraordinary Situation

The question of "adoption for birth" (or embryo adoption) is one of the most controversial issues in contemporary Catholic bioethics. For more than two decades, it has generated serious and legitimate debate among theologians, bioethicists, and Church authorities. My own position, reflected in this endorsement and in line with the entire Magisterium of Pope Francis, has consistently been that embryo adoption cannot be understood as a new or alternative form of procreation, nor as a justification for IVF. Rather, it can only be considered, under very strict conditions, as an extraordinary and provisional remedy in a situation of radical abandonment. Etheredge's book moves in this same direction, while offering a distinctive theological depth. The adoption of an abandoned embryo is interpreted as an analogical participation in the logic of the Incarnation itself: a free and gratuitous act of welcoming a vulnerable human life that one did not generate, but for which one assumes responsibility. This point is crucial. Embryo adoption does not legitimize the practices that produced the injustice; on the contrary, it presupposes their moral rejection. It makes sense only if accompanied by a clear and uncompromising opposition to the production and cryopreservation of embryos. Understood in this way, adoption for birth is not oriented towards fulfilling the desires of adults, but toward protecting the fundamental right to life of a concrete human being who otherwise has no future.

Theological Insight: The Annunciation and the Meaning of Motherhood

One of the most original contributions of Etheredge's work is his theological meditation on the Annunciation as a key to understanding the moral significance of welcoming embryonic life. Mary is not presented merely as a biological model, but as an anthropological and ecclesial icon. Her "yes" is not an act of possession or production, but of reception. Motherhood, in this light, is revealed as hospitality to life, not mastery over it. This distinction allows Etheredge to avoid superficial analogies between embryo adoption and surrogacy. The latter is grounded in contractual logic and instrumentalization, while the former, when properly understood, is grounded in gratuitous care for an abandoned life. This perspective also has ecclesial implications. The Church cannot limit herself to condemning injustice at the level of principles alone. She is also called to discern how to witness to the Gospel of life within the tragic fractures of history, without compromising truth but also without abandoning concrete persons.

Toward an Integral Bioethics

From a bioethical standpoint, one of the greatest merits of this book is its refusal of two opposite but equally inadequate approaches: abstract legalism and sentimental pragmatism. Abstract legalism risks transforming moral principles into barriers that prevent any engagement with real suffering. Sentimental pragmatism, on the other hand, risks dissolving moral truth in emotional reactions.

Etheredge avoids both by proposing what can rightly be called an integral bioethics, one that holds together truth, justice, mercy, and responsibility. The tragedy of frozen embryos forces us to think ethically in "limit situations," where the application of moral principles requires discernment rather than mechanical deduction. This does not mean relativizing moral norms, but recognizing that moral responsibility is always exercised in history, not in abstraction.

Why This Book Matters Today

The Salvation of the Embryonic Person is a challenging book, intellectually, morally, and spiritually. It unsettles comfortable positions and forces the reader to confront the deepest questions about human dignity, technological power, and moral responsibility. It does not offer easy answers, nor does it claim to close the debate. Instead, it elevates the discussion, grounding it in a profound anthropological and theological vision. At a time when embryonic human beings risk becoming invisible casualties of technological progress, this book restores their face, voice, and moral claim.

Conclusion

I strongly recommend this book to bioethicists, theologians, physicians, jurists, and all those who are committed to thinking seriously and responsibly about the beginning of human life. It is a work that does not evade complexity, but neither does it surrender to despair. Ultimately, how a society treats its most vulnerable members, especially those at the very beginning of life, reveals not only its level of technological development, but its moral identity.

Introduction

Can the Contemplation of the Annunciation Reverse the Plight of Embryonic Children[1]

The salvation of all, then, begins with the salvation of each one of us at conception. This does not preclude, but requires baptism and the Christian life, nevertheless it is the beginning which begets the salvation of us all. Thus the salvation of the least of us is for the benefit of us all![2]

This book, then, does not ignore the plight of sinners, of mankind, of many others, almost too numerous to count, whether marginalized, migrants, displaced people, addicts, those trafficked, those at war, or indeed any in any or every kind of human need or crisis. Rather, this book simply adopts, literally, a focus on our embryonic beginnings in the hope against hope (cf. Rom 4: 18) that they, along with the rest of us, will be included in a new spring-time of God's overflowing, everlasting love. God, having thought of us from all eternity, does not create us and set us sail, as it were, on a stormy sea; but, rather, accompanies us in the very nature of how we are brought

---

[1] The seeds of this very different book, as well as some of the writing, were published in a chapter under the title of "The Annunciation and Embryo Adoption", in the following work: *Human Embryo Adoption, vol. 2, Catholic Arguments For and Against*: https://www.ncbcenter.org/store/human-embryo-adoption-vol-2-catholic-arguments-for-and-against.

[2] I am very grateful to Dr. Anthony Williams for his careful proofreading; and, therefore, any mistakes or errors that remain are mine.

to exist. And so, having thought of us from all eternity, God rains His grace like water on a mountain, giving birth to springs of living water (cf. Jn 7: 38) that rise up, as at *Baptism*, and overflow through all the other sacraments. But why begin with the unborn? Because this is where God begins! Read on!

Just as Pope Leo XIV says that if we cannot love our neighbour who we do see, we cannot claim to love God who we do not see, so if we do not see that our salvation is linked to the salvation of all[3], we risk being "unsaved". Indeed, as Pope Leo says early on: 'contact with those who have neither power nor greatness is a fundamental way of encountering the Lord of history'[4]; and, therefore who, more than human embryonic children, 'have neither power nor greatness'! And, if the 'poor … are the very flesh of Christ and not just a sociological category (cf. Dilexi Te, 110)'[5], then let us not forget to meet Him in the person about to be aborted or frozen in that who can be more lonely than them[6]. Similarly, if peace comes from 'announcing the presence of a defenseless God, in whom humanity can

[3] Cf. Pope Leo XIV, *Dilexi Te*: "On the Love for the Poor", eg. paragraph 118 but really the whole document argues this: https://vatican.va/content/leo-xiv/fr/apost_exhortations/documents/20251004-dilexi-te.html.

[4] *Dilexi Te*, paragraph 5.

[5] Zenit Staff, 16th of November, 2025, "The end of time and the drama of loneliness that permeates all forms of poverty, according to Pope Leo XIV": https://zenit.org/2025/11/16/the-end-of-time-and-the-drama-of-loneliness-that-permeates-all-forms-of-poverty-according-to-pope-leo-xiv/?eti=28727.

[6] Zenit Staff, 16th of November, 2025, "The end of time and the drama of loneliness that permeates all forms of poverty, according to Pope Leo XIV".

discover itself as loved only by caring for him (cf.*Lk*2:13-14)'[7], perhaps this invites us to see anew that caring for embryonic human life is an opportunity to rediscover God's love of us. As Dr. William Lile says, so succinctly, "A patient is a person, no matter how small"[8].

So, as Pope Leo goes on to say elsewhere, we need to look up from our own problems and seek 'the *grace of an overarching vision*, a perspective capable of grasping the horizon, of looking beyond'[9]. And, therefore, while this little book focuses on the unborn, I hope it opens out upon the vista of the Lord's plan, from everlasting to everlasting, of our eternal salvation. And, what is more, although we cannot expect from the word of God what is derived from other disciplines, nevertheless, the God-inspired-word possesses a certain powerful creativity: 'Where educational communities allow themselves to be guided by the word of Christ, … .They respond with creativity, opening up new possibilities for the transmission of knowledge and meaning in … research, because the Gospel does not grow old but makes "all things new" (*Rev* 21:5)'[10]. Just, then, 'as there

---

[7] Message of the Holy Father Leo XIV for the 59th World Day of Peace (1 January 2026), 18.12.2025: https://press.vatican.va/content/ salastampa/en/bollettino/pubblico/2025/12/18/251218a.html.

[8] "Pro-Life Doc: Where Science and Scripture Meet to Defend Life": https://www.prolifedoc.org/.

[9] Pope Leo XIV, 27th October, 2025, « Eucharistic Celebration with Students from the Pontifical Universities »: https://www.vatican.va/content/leo-xiv/en/homilies/2025/documents/ 20251027-messa-univ-pontificie.html.

[10] 27th October, 2025, "Apostolic Letter DRAWING NEW MAPS OF HOPE of Pope Leo XIV on the occasion of the 60th Anniversary of the Conciliar Declaration *Gravissimum educationis*, paragraph 1: 1:

is no faculty that faith does not illuminate, so there is no science that theology can ignore. Through a well-rounded study, you are therefore called to offer your precious contribution to the discernment and resolution of the challenges that confront both the Church and humanity as a whole'[11]. Therefore, as we shall see, there is a dynamic interaction, in the course of this short book, between embryology and theology.

Holiness comes to meet the Sacredness of Human Life

Let us reflect, then, that if we have contact with embryonic human children, we are 'encountering the Lord of history' who was Himself, as indeed was His Holy Mother, once an *embryonic child.* Furthermore if, as St. Paul VI said: 'Human life is sacred – all men must recognize that fact', Our Predecessor Pope John XXIII recalled, 'from its very inception it betrays the creating hand of God'[12] – then not only are we investigating the sacredness of human life but the extraordinary sanctity of Our Blessed Lord and His Mother Mary! Thus, if on the one hand we are talking about the transmission of human life from our parents then, on the other hand, we are also talking about another mystery: that in a mysterious way there is the

---

https://www.vatican.va/content/leo-xiv/en/apost_letters/documents/20251027-disegnare-nuove-mappe.html.

[11] Pope Leo XIV, "Audience with the members of the International Theological Commission, November 26, 2025": https://press.vatican.va/content/salastampa/it/bollettino/pubblico/2025/11/26/0908/01646.html.

[12] *Humanae Vitae*, 13: https://www.vatican.va/content/paul-vi/en/encyclicals/documents/hf_p-vi_enc_25071968_humanae-vitae.html.

"transmission" of "sacredness", from the very 'creating hand of God', to each and every embryonic human child. Indeed, given the holiness of God Himself, it is clear that the action of God expresses the sacredness of God Himself. Thus, can an action of God, expressing the sacredness of God, act without imparting a "touch" of grace? And, if grace always requires a personal subject, then we can say that this initial, human sacredness, is a sign that the whole person was present from the first instant of human fertilization.

As this book develops, we might glimpse how the sin of Adam, original sin, that transmission of the defect of original grace "meets", as it were, the contrary movement of the welcoming grace of God!

And if it is true, that 'That which was not assumed has not been healed; but that which is united to God, the same is saved'[13], then it follows that Christ took on the very first instant of human conception, both making it His own and, therefore, redeeming us from the very first instant of human conception – even if the details of His conception remain shrouded in mystery. As Pope Leo XIVth says: 'If Jesus has a real body, the communication of the truth of God is realized in that body, with its own way of perceiving and feeling reality, with its own way of inhabiting and passing through the world'[14]. Furthermore, 'by His incarnation the Son of God has united Himself in some fashion with every man' (*Gaudium et Spes*,

---

[13] *The Faith of the Early Fathers*, vol 2, passages selected and translated by William Jurgens, (Collegeville, Minnesota: The Liturgical Press, 1979), p. 41.

[14] "Full text: Pope Leo XIV's general audience given Jan. 21, 2026": https://www.osvnews.com/full-text-pope-leo-xivs-jan-21-2026-general-audience/.

22); and, therefore, as Christ 'united Himself … with every man', so the Son of God has redeemed us all from the first instant of being a man, male and female, namely the first instant of human conception: the first instant of the fertilization of an egg-cell by a sperm. Thus for the *Incarnation* to unite us, as it were, with Christ, each of us needs to be a whole human person even if, as the Psalmist David says, we are an 'unfinished vessel' (Ps. 139: 16) – we are nevertheless a personal subject.

Each act of God creating a human person "echoes" or recapitulates the beginning

Indeed, although the subject of the opening pages of Genesis is the creation of the whole universe, what is striking is that the account begins with what is almost indiscernible, namely, the 'earth [which] was without form and void' (1: 2). This is a remarkably poignant beginning which shows a grasp of what is almost unimaginable: that out of an almost contradictory beginning of almost nothing everything emerges from a sequence of successive acts of God. Thus, as we see later in Scripture, there is a prescient similarity in David's understanding of his conception. He says in the psalm: 'Thy eyes beheld my unformed substance' (Ps 139: 16)[15]. And, indeed, the Hebrew word is '*golmi*', which brings together two images, 'unfinished' and 'vessel'[16]. But not only does it bring together these

---

[15] Cf. also Etheredge, *Conception: An Icon of the Beginning*: https://enroutebooksandmedia.com/conception/.

[16] F. Brown, et al, *A Hebrew and English Lexicon of the Old Testament*, Oxford: University Press, 1939, page 166.

two images, indicating an actual beginning which is yet 'unfinished', and a beginning to what is bearing an interiority, as a 'vessel', but that also the word ends with a personal ending: "Your eyes beheld *my*[*] unformed substance" (Ps 139: 16)'[17]. Again, David echoes the very personal act characteristic of the second account of creation in the *Book of Genesis*. In other words, it is the whole which is 'my unformed substance'[18]: I begin existence as a whole human person which is dynamically orientated to the manifestation of what is there from the beginning, namely, the whole "I" or "you". Furthermore, St. Paul develops this usage when, speaking of the Christian life within us, he says: 'we have this treasure in earthen vessels' (2 Cor 4: 7)[19]. Mary, then, is the perfect fulfilment of the human person; indeed, both in herself and as a being-in-relationship. For grace builds upon nature, which is that we are in-relationship to one another, and Mary is the dwelling "person-place" of the Holy Spirit.

---

[*] This signifies an emphasis added to the quoted text.

[17] *The Holy Bible*, Revised Standard Edition, Catholic Edition, San Francisco: Ignatius Press, 1966, p. 569.

[18] Cf. Etheredge, "Your eyes beheld my unformed substance" (Ps 139: 16): The Unspeakable Crime of Abortion (Evangelium Vitae, 58)", Chapter Eleven of *Scripture: A Unique Word*: https://www.cambridgescholars.com/product/978-1-4438-6044-4.

[19] Further discussion of this imagery takes us too far from our subject; but, suffice it to say, that this concept of 'my unfinished vessel' is both a unique word in Scripture and apt for expressing the beginning of each one of us.

The tragic nature of the predicament of embryonic and developing children

> Although in a different context, Pope Leo XIVth, as we see, recognizes how what is natural in itself, for married people to want to become parents, is commercialized as the child is commodified: 'Around us, a distorted economy tries to profit from everything. We see how the marketplace can turn human yearnings of seeking, traveling and beginning again into a mere business. Let us ask ourselves: has the Jubilee taught us to flee from this type of efficiency that reduces everything to a product and human beings to consumers?'[20]

On the one hand there are those obliterated by abortion, for whom now is too late and from whom we need to beg their forgiveness and intercession for the rest of us; albeit, however, they are saved in ways known only to God, for not only had they no personal sin, but that they were the object of a terrible attack on human life. Moreover, they were both united to Christ at His *Incarnation* and could not be saved nor intercede for us if this were not so. And, on the other hand, there are those willing mothers, willing even to lose their own life so that their child may be born and, indeed, may all mothers intercede for us as we continually struggle to do good[21].

---

[20] Zenit Staff, January 6th, 2026, "Is there life in our Church? What has the Jubilee taught us?": https://zenit.org/2026/01/06/is-there-life-in-our-church-what-has-the-jubilee-taught-us/?eti=30367.

[21] Cf. Abigail Jorgensen, March 14th, 2024, "St. Gianna Beretta Molla: A Saint for Our Times": 'She was canonized for her heroism in being

Furthermore, there are those women who, coming out of the abortion industry itself, give witness to their experience seeing, as they do, the tragic reality of both abortion, the aftermath, and the effect on the staff of a culture of drinking and, therefore, the need for prayers for all involved. At the same time, the woman who has aborted her child needs to be helped; and, while this book is not about her suffering, we nevertheless have to remember her in our prayers and, when possible, speak with sensitivity to her. For you or me may be the person who releases her from the 'personal prison of shame'[22].

But, having said all this, the question posed by this book is whether or not this different angle, this different focus, this different way of coming upon the question of the plight of embryonic children, will help us to hope against hope that God will act in this generation to spare the millions who are being slaughtered. And they are slaughtered in the dark, as it were, without seeing the wonderful gift of creation in which, as gifts themselves, they are brought to exist. And, in short, this is the answer that this book is gravitating to: that we need to see more clearly that the life of the unborn is inseparable from the mystery of the *Immaculate Conception*, the *Incarnation*, and the salvific love that Christ has for each one of us. Our salvation is inseparably linked to the salvation of the least and poorest

---

willing to die for her child': https://churchlifejournal.nd.edu/articles/st-gianna-beretta-molla-a-saint-for-our-times/

[22] Heather Lawson, February 10th, 2026, "Why it took me more than 20 years to share my abortion story." https://www.christianpost.com/voices/why-it-took-me-more-than-20-years-to-share-my-abortion-story.html; and cf. my own experience of losing a child to abortion in a piece called "Indelible" in *The Prayerful Kiss*: https://enroutebooksandmedia.com/theprayerfulkiss/.

of all! Those that have neither 'power nor greatness' to deter others from their abuse or destruction.

Where to begin, if not with the obvious signs of the obliteration of children by abortion and, if they survive, being left to die? And with the experimental mining of the flesh of human embryonic children, their extermination or, if frozen, their melting, as if they have an expiry date beyond which they are to be wasted. Alternatively, even if a child is transferred to the womb of a waiting woman, there is no guarantee that, even then, that child will not be deleted – especially if there are numerous children transplanted into the one woman's womb. Albeit we now know that a child can be born after having been frozen, thawed, and transferred to a mothering woman after being frozen for thirty-one years![23] So, yes, we are addressing the needs of actual children – however they are described. Otherwise there is no point to this immense expense if all it involves is the making and unmaking of indifferent human cells.

But, more widely, is the relentless medicalization of the human body inuring us to what it means: that each one of us is a gift to be given and received in love? Is the very "production" of fertilization in a glass dish already distancing us from the human reality that, even in the midst of the mechanization of his or her conception, that this is still a child? Otherwise, as I have said, there would be no point

[23] By Julio Tudela and Cristina Castillo of the Bioethics Observatory, Published On: September 16th, "Baby born after 31 years of embryonic cryopreservation": https://bioethicsobservatory.org/2025/09/baby-born-after-31-years-of-embryonic-cryopreservation/48072/

to these procedures if we are not addressing, actually, the life-start of each one of us. However, unlike the rest of us, these children are at the merciless end of a production process which may or may not lead to their right to completing human development being recognized and facilitated? And if the child is not deleted as deemed "in excess" of who is wanted, he or she is then frozen in liquid nitrogen and remains in an uncertain state for an indeterminate time. So we begin this short book with a social sketch, as it were, as to how technology has been harnessed so as to shackle the unborn. And, at the same time, what is the 'good' and the 'light' that are evident in our times? But lest we forget, let us remember that it is precisely Christ and His Church that has come to rescue us that we might extend His welcoming embrace to all.

While the love of God and life are the greatest motives for protecting human beings, let us not forget that the 'fear of God is the beginning of wisdom' (Proverbs 9: 10). Thus St. Paul addresses the seriousness of destroying human life:

> "Do you not know that you are God's temple and that God's Spirit dwells in you? If anyone destroys God's temple, God will destroy him. For God's temple is holy, and that temple you are' (1 Corinthians 3: 16-17, Catholic RSV).

In answer to the objection that St. Paul's address seems to be limited to protecting the existence of Christians, one in whom the Spirit of God dwells, it can be said that God has already defended human life as a whole:

> 'Whoever sheds the blood of man, by man shall his blood be shed; for God made man in his own image' (Gen 9: 6). This verse again emphasized the sacredness of human life; this time, because each of us is an "icon" of God. And St. John Paul II, too, drew on Scripture to defend human life, seeking the roots of human violence and offering, on the contrary, a Gospel of Life[24].

And, as Pope Leo XIVth says, 'Since God has become one of us, every human creature is a reflection of him, bearing his image and containing a spark of his light. This calls us to recognize the inviolable dignity of every person and to offer ourselves in mutual love for one another. Moreover, the Incarnation demands a concrete commitment to the promotion of fraternity and communion. Through this commitment, solidarity becomes the criterion of all human relationships, calling us to strive for justice and peace, to care for the most fragile, and to defend the weak'[25].

And who or what is more fragile and weak than human embryonic life?

---

[24] Cf. Chapter 1 of The Gospel of Life: 'CHAPTER I - THE VOICE OF YOUR BROTHER'S BLOOD CRIES TO ME FROM THE GROUND': https://www.vatican.va/content/john-paul-ii/en/encyclicals/documents/hf_jp-ii_enc_25031995_evangelium-vitae.html.

[25] Zenit Staff, January 4th, 2026, "Pope Leo XIV: a call to examine our spirituality and how we express it": https://zenit.org/2026/01/04/pope-leo-xiv-a-call-to-examine-our-spirituality-and-how-we-express-it/

In what follows I range over the times in which we live, the challenge of defining the terms in use, the ordinary and the extraordinary experiences that the Creator has revealed and on which we are to ponder, particularly Mary's own *Immaculate Conception* and the *Incarnation* of the Son of God. While it is relevant to emphasize the gravity of destroying human life, and the gift of repentance and forgiveness for anyone who turns to the 'ample arms'[26] of Jesus Christ, what particularly concerns us here is God's merciful embrace of those who, utterly unjustly and without warrant, are either extinguished or torn apart. Thus there will be a discussion of what is relevant in terms of embryology, theology, Scripture and the progress to be made of grasping what it means to say that God can save the very least among us outside of His own sacraments.

Three kinds of published work

The following titles are now on Amazon, in production or, as with this book, in the process of being written; but I list them here in three groups: Scripture, theology, and philosophy; Bioethics; and autobiographical works; indeed, the third group began with a prayer in front of the Lady Chapel at our local Catholic Church, in the hope of writing a more widely accessible read.

Scripture: A Unique Word (2014); A trilogy on Faith and Reason called, From Truth and truth: Volume I-Faithful Reason (2016),

---

[26] Pope Francis' Letter on Dante, *Cantor Lucis Aeternae*, paragraph 5, quoting from Dante, Purg. III, 118-123: https://press.vatican.va/content/salastampa/it/bollettino/pubblico/2021/03/25 /0181/00393.html#ing.

Volume II-Faith and Reason in Dialogue (2016), Volume III-Faith is Married Reason (2016).

A second section of work, then, focuses on primarily bioethical questions. The Human Person: A Bioethical Word (2017); Conception: An Icon of the Beginning (2019); Mary and Bioethics: An Exploration (2020); Reaching for the Resurrection: A Pastoral Bioethics (2020); and Unfolding a Post-Roe World (2022); and Human Nature: Moral Norm (2023). Furthermore, the *National Catholic Bioethics Centre* of America invited a contribution to Volume Two of a book on *Human Embryo Adoption: Catholic Arguments for and against*; my contribution is called "The Annunciation and Embryo Adoption" and argues that embryo adoption is a Marian participation in the saving work of Jesus Christ (2025). While the following two books have autobiographical elements, as do most other books, they are more about the situation of our times than personal experience and, therefore, they warrant being in this section. The Word in Your Heart: Mary, Youth, and Mental Health (2024); Transgenderism: A Question of Identity (2025), which ranges over our society and comes to a gradual focus on transgenderism which had turned up, little by little, in previous books. Francis Etheredge, together with Elizabeth Rex, co-authored The Zygote of Christ and the Mystery of Man (2025). And now there is Transgenderism Two: Evidence and Identity (2026), which examines more deeply the recent documentation from the Supreme Court of the United Kingdom, a Report on Paediatric Medicine from America, and a Family Court Judgement in Australia, along with other material. A Shorter Bioethics is a summary statement of the major work which has been published up to now in this collection (2025). Now, of course, there is this one: The Salvation of the Embryonic Person: Beginning with the Incarnation and the Immaculate Conception (2026).

The third section of work is much more autobiographical and covers a variety of life events: The Family on Pilgrimage: God Leads Through Dead Ends (2018); The Prayerful Kiss: A Collection of Poetry and Prose (2019); Honest Rust and Gold: A Second Collection of Prose and Poetry (2020); Within Reach of You: A Book of Prose and Prayers (2021); Lord, Do You Mean Me? A father-Catechist!" (2023); An Unlikely Gardener: A Book of Prose and Poetry (2023); and now Time! Ladies and Gentlemen – Thoughts on Last Things In Response to James Sale's Trilogy: HellWard, StairWell, and DoorWay (2025); God is in the Detail: Pilgrimage, (2026).

Chapter One

A "dramatic struggle between good and evil, between light and darkness"

"Therefore man is split within himself. As a result, all of human life, whether individual or collective, shows itself to be a dramatic struggle between good and evil, between light and darkness. Indeed, man finds that by himself he is incapable of battling the assaults of evil successfully, so that everyone feels as though he is bound by chains. But the Lord Himself came to free and strengthen man, renewing him inwardly and casting out that "prince of this world" (John 12:31) who held him in the bondage of sin. For sin has diminished man, blocking his path to fulfilment" (*Gaudium et Spes*, 13)[1].

How does the struggle between 'good and evil, between light and darkness' show itself?

We live in a society in which there is a lack of regulation as regards what can and cannot be done to human beings, whether at conception or at any other point in life. Although there are wonderful declarations of human rights, they almost all seem completely impotent when confronted with the inroads of businesses which are based on the exploitation of human subjects. The binding agreement of Beijing, that there are two sexes, male and female, is almost totally ignored by the very institution where it was recognized, namely, the

[1] No italics in the original quotation; but they are used here to distinguish the quotation as an opening text.

United Nations. Therefore there is not only the rampant destruction, experimentation, and freezing of real embryonic human beings, there is the spreading mentality that the human being is a material for "manipulation", including the pretence of changing a person's sex.

What does this mean for our subject of the most vulnerable, the unborn, the "farmed" human beings, the ones deleted, experimented upon, trashed and frozen? It means that, in truth, there is almost no prohibition on the "mining" of the human, embryonic child. Clearly all kinds of emotional arguments are advanced, namely, that fertilizing a child in a glass dish gives an infertile woman an opportunity to have a wanted child. However, there is now a move to develop a truly investigative approach to infertility, resulting in the uncovering of a range of factors obstructing health and prohibiting conception. This type of medicine, obviously compatible with the best methods of natural family planning, is called 'Restorative Reproductive Medicine':

> 'RRM prioritizes sound medical approaches to optimize patients' natural bodily function and health; optimizing health and natural function in the least invasive manner possible is the clear preferred approach supported by clinical evidence'[2].

---

[2] To engage more fully with this development, go to the website and discover more: https://iirrm.org/the-progress-and-promise-of-restorative-reproductive-medicine/?mc_cid=fb13474bd2&mc_eid=b1f4a58a67; and also cf. Kristen Curran, January 13th, 2026, "The four key takeaways on the restorative approach to infertility treatment, from the largest NaPro

But, more generally, what help is given to women to naturally investigate their infertility, to see whether it is related to prior contraceptive use, previous abortions, or other causes more discernible through natural family planning methods?[3] There is, too, however, the possibility that the men is the one who is infertile or even both the man and the woman. Therefore a discerning investigation is indeed necessary for both of them.

Or what of the view that at the beginning of life there is only a "blob of cells". And if each one of us began as a "blob of cells", then that "blob of cells" was wonderfully capable of manifesting me, and you. What is more, if cultivating that "blob of cells" has become a billion, if not a trillion-dollar industry, precisely because that "blob of cells" is at the origin of each one of us, then clearly there is more to that "blob of cells" than is admitted by the very industry which exploits it.

We saw this beginning with the exploitation of women in the infamous experiments which led to the development of the contraceptive pill: a pill which unnecessarily medicalizes the natural cycle of fertility and infertility in the woman; and, indeed, has gone on to be an almost "automatic" response to the onset of puberty in girls. Not only does the contraceptive pill mask a woman's period but it

---

study to date: The 2025 study analyzed data from 1,310 couples at the Fertilitas Center in Madrid": https://naturalwomanhood.org/largest-napro-study-infertility-treatment-rrm/.

[3] Etheredge, *Human Nature: Moral Norm*, especially where the benefit of natural family planning methods are used diagnostically, both recognizing ovulation as an indicator of a woman's general health and as a method of identifying and remedying other problems: https://www.amazon.co.uk/Human-Nature-Moral-Francis-Etheredge/dp/1956715517.

hides the vital signs of her health shown by the timeliness of her period and any problems with it[4]. Now we have the deliberate destruction of the unborn, whether by pill, surgical or other methods, the prevalence of doubtful methods of determining the end of life, known as "brain death", which in practice facilitates organ stripping. In addition, since 2020, 54,000 women have been 'admitted to NHS hospitals for treatment following complications linked to abortion pills'[5]. What is amazing is that there is no liability for this burdensome pharmaceutical experiment on women which is making money out of the death of children, harm to women, and treatments that are only necessary because of disregarding life. There is, too, the practice of the deliberate killing of the infirm, elderly, or mentally ill, by various means[6]. Now we see that growing up has been transformed into a traumatic event by medical interference in puberty, or even younger, from being a process that a young man or a young woman ordinarily went through. After all, adolescence is, in general, a time when children and adolescents need help to discover who they are and what they want to do with their lives. But, owing to gender ideology, adolescence is now a time that is fraught with the dangers of a child and his or her parents being seduced into thinking that the young person is in the "wrong body" and, as a consequence,

---

[4] See Etheredge, *Human Nature: Moral Norm.*

[5] October 13th, 2025, "54,000 hospital admissions as a result of pills-by-post": https://spuc.org.uk/54000-hospital-admissions-as-a-result-of-pills-by-post

[6] Cf. Etheredge, *Reaching for the Resurrection: A Pastoral Bioethics*: https://enroutebooksandmedia.com/reachingfortheresurrection/.

losing integral, natural, essential body parts of the boy or the girl that he or she is, actually and irrevocably[7].

In this context of the sweeping medicalization of the body, which includes anti-depressants, anti-anxiety tablets, anti-almost any uncomfortable feeling, rather than recognizing their human and personal significance, there is an almost unbearable silence owing to the stifling of human suffering.

The reality of an appalling death toll that appears to be constantly multiplying

What does not need medicalizing? In this era there seems to be no end for answering everything with a drug, a treatment, or surgery. There is the development of technological "remedies" for infertility which completely circumvent the investigation of that infertility and the possibility of remedying it, either through natural family planning expertise or other, actually therapeutic medical methods. Thus the very origin of what is called *in-vitro* fertilization, whereby a woman's infertility was circumvented[8] in order to provide a pretext for conducting experiments on the beginning of human life, is constantly being perpetuated. However, this does not reflect on the children conceived; rather, it reflects on the unmitigated

---

[7] Cf. Etheredge, *Transgenderism: A Question of Identity*: https://enroutebooksandmedia.com/transgenderism/.

[8] Cf. Etheredge, "On Regulating IVF", in Ethics & Medics, Volume 41, Issue 7, July 2016: https://www.pdcnet.org/852585 F900588A85/file/5A9F 94ADC0FC0B1D852586630061A719/$FILE/em_2016_0041_0007_0001_0002.pdf.

willingness of so-called scientists to pursue any avenue of action without reference to the truth of what they are doing or the reality of what is good, either for themselves or those they purport to help. Indeed, to put this in some kind of perspective, in a report on IVF in 2016 it was said that '5 million children' were fertilized using IVF; and, to give a figure to the increase in human embryos in American freezers, it is about '19, 000' annually, while in the UK the increase in frozen embryo is '45, 000' per year[9]. In a more recent report it is claimed that there are '"13–17 million infants" born via IVF since 1978'. Thus if the number of children born from an IVF fertilization has possibly trebled, then it is estimated that 'The *in vitro* fertilization business has likely directly killed 270 million unborn children or more since the beginning of the process in 1978'[10].

What is more, abortion and abortifacient contraceptives multiply that number exponentially. Thus frozen children, then, exist against a background of the devastating destruction of human life; and, therefore, are a continual reminder of the commodification of children. In an all too literal expression of the commodification of children is the company that, literally, turns human embryos into jewellery: "We specialise in crafting unique and meaningful

---

[9] Justo Aznar-Lucea, Miriam Martínez-Peris, and Pedro Navarro-Illana, "FROZEN EMBRYO ADOPTION", in THERAPEÍA 8 [Julio 2016], 103-120, ISSN: 1889-6111, p. 106: https://www.academia.edu/30973678/FROZEN_EMBRYO_ADOPTION

[10] Matt Lamb, August 14th, 2025, "IVF has killed more than 250 million human embryos since 1978: analysis": https://www.lifesitenews.com/analysis/ivf-has-killed-more-than-250-million-human-embryos-since-1978-analysis/?utm_source=daily-usa-2025-08-14&utm_medium=email.

keepsakes using unused IVF embryos"[11]. If it is still asserted that at the beginning of human life there is only a "blob of cells", then the industrial destruction of those cells has assumed such giant proportions that this clearly contradicts that claim. In other words, the very wholesale slaughter and freezing of what is euphemistically called a "blob of cells" is tantamount to evidence that these "blobs of cells" are undoubtedly children paying the price of the commercialization of the life and death of our most vulnerable brothers and sisters. Otherwise, why would there be industries destroying, or bringing to exist, a mere "blob of cells"?

Thus the more the plight of the aborted and frozen unborn is pondered, the clearer it becomes that there is a new context for the discussion of the good that is to be done for those who suffer the following, triple injustice. Firstly, that a child is conceived outside of the marital act. Secondly, that a child is to be part of a "quality control procedure", applicable to objects but not to persons, which may entail reductive de-selection or being aborted, being discarded as waste, or being experimented upon. And, thirdly, the possibility that the child will be frozen, interminably, with no certainty of either being thawed, surviving the thaw, or being able to complete the natural unfolding of the manifestation of the human person begun at conception.

So we are in the midst of a great slaughter and, it seems, few of us have eyes to "see" it. Or, if we can see the human catastrophe, the

---

[11] Alyssa Murphy, November 4th, 2025, "Embryo Jewelry? Catholics React to Frozen Children Encased in Necklaces and Rings": https://www.ncregister.com/blog/embryo-jewelry-frozen-children-encased

possibility of reversing this tide appears as fragile as the slightest expression of an embryonic child's life surviving the terrible onslaught of wave upon wave of their destruction and commodification. Countless sons and daughters, brothers and sisters, nieces and nephews, granddaughters and grandsons have been lost without notice – except that there is a trace, like a scar, that runs through the hearts of those who know their child has irrevocably passed and cannot forget. What has happened to harden the hearts of so many, so that so many are heartlessly commodified and cancelled?

Do we not know God loved us into existence – having thought of us from all eternity? And that even if the circumstances of our life were difficult and uncertain, did we not know that rubbing out the existence of a child will leave an indelible smudge on our own heart – aching after that relationship which came to exist and, while it remained, pleaded for however long? But, safe in the hands of God, all those who lost their lives tragically are there, awaiting the possibility of reconciliation. Thus people chained to what they have done can be released by forgiveness and reconciliation.

A general comment on the new anthropological context of marriage and procreation: The Person-as-Gift

Some of us live in a very technocratic age where we marvel at what is possible, technically, from heart operations to buildings; and, in a sense, this is right and just as it gives glory to God for the good use of His gifts. However, this kind of thinking can be very hollow and heartless when it is applied to human beings. Thus, for example, an eye can be described in terms of its function of seeing, as if seeing is simply the passage of light through the pupil, falling as it does on the retina, and impressing an image upon our brain through the use of one of the most concentrated bundles of nerves in the body. And so we can see a tendency to see the 'human person as a material object subject to the laws of mechanics' or, similarly, to being a 'complex informational system'[12]. Either way, instead of looking at science in the light of the human person, the human person is looked at reductively, as if 'mechanics' and an 'informational system', which are products of human intelligence[13], are an adequate account of the human person who is created out of love by God. But not just created by God but created in the 'image and likeness' of God (cf. Gn. 1: 26-27).

Sight, however, is an expression of the whole person and is, in a certain way, expressive of what is in the heart of a person. A heart dominated by lust sees what we should be ashamed to admit to seeing. And yet this is the mentality that has destroyed so many hearts

---

[12] Richard Spinello, Summer 2018, *National Catholic Bioethical Quarterly*, of "Bioethics and the Human Soul: Pope St. John Paul II's Reflections on Ensoulment", pp. 291-292: https://www.pdcnet.org/C1257D43006C9AB1/file/1BCFFCF397DE7BAD852582F8004847AC/$FILE/ncbq_2018_0018_0002_0095_0120.pdf.

[13] An expression used by *Antiqua et Nova*, to express the relationship between artificial intelligence and human intelligence: https://www.vatican.va/roman_curia/congregations/cfaith/documents/rc_ddf_doc_20250128_antiqua-et-nova_en.html. The full quote is: 'Considering all these points, as Pope Francis observes, "the very use of the word 'intelligence'" in connection with AI "can prove misleading"[69] and risks overlooking what is most precious in the human person. In light of this, AI should not be seen as *an artificial form* of human intelligence but as *a product* of it' (paragraph 35).

and marriages, as pornography contributes to marital breakdown, along with a general promiscuity, as the tendency to delay or even avoid marriage altogether impacts the possibility of welcoming children. Indeed, it is as if the hearth, the heart of the family, which is intended for the welcoming of children, has grown cold: a cold literally expressed in the willingness to both abort or to freeze embryonic children.

But a heart, expressing wonder, the wonder of seeking to understand what we see, whether in nature or in ourselves, is a heart open to "what is" and the possibility of a path of truth leading to God. Thus, beginning with the *Second Vatican Council's* document on the Church, *Lumen Gentium*, we are reminded that married couples 'signify and partake of the mystery of that unity and fruitful love which exists between Christ and His Church' (11). Even more recently, the Church says that 'Jesus restores validity to God's original plan, going beyond the norm given by Moses and recalling an older one, while at the same time underlining a divine presence at the very root of this relationship: "What God has joined together, let no man separate" (*Mt* 19:6)'[14]. But returning to the Council, the Fathers went on to revive an ancient definition of the family as the 'domestic Church'[15]. Developing their thought further, in *Gaudium et Spes*, the Fathers of the Council spoke of marriage as 'the conjugal covenant

[14] *Una Caro*, paragraph 27: https://www.vatican.va/roman_curia/congregations/cfaith/documents/rc_ddf_doc_20251125_una-caro_it.html.

[15] Paragraph 11: https://www.vatican.va/archive/hist_councils/ii_vatican_council/documents/vat-ii_const_19641121_lumen-gentium_en.html.

of irrevocable personal consent'[16], enriching our understanding with a biblical concept that expresses the relationship of God to creation, to His people, to the mystery of marriage taken up by the prophet *Malachi* (2: 14), and to Christ Himself in St. Paul's *Letter to the Ephesians* (5: 32).Thus the covenant permeates marriage, as it were, with the action and promise of God; and, as such, really roots it in the mystery of Christ's love of His Church which is witnessed, very concretely, in Christ granting His mother's request to give the newly married couple His new and better wine (Jn. 2: 1-12).

There has been, however, a tendency to equate marriage, particularly the marital act, to the behaviour of animals: as if husband and wife are almost "animal-like" in the inclination to come together and to have and to raise children. On the one hand this view entailed a certain naturalistic understanding of the action of the spouses; but, on the other hand, it almost seemed to signify an almost less than human mentality. With the reflection of Pope St. Paul VI, building on the work of the aforementioned *Second Vatican Council*, we see a new, personalist account that marriage is both a ministry of cooperation with God (cf. *Humanae Vitae*, 13, 8) and that spousal union is at once the expression of an 'inseparable connection, established by God, which man on his own initiative may not break, between the unitive significance and the procreative significance which are both inherent to the marriage act' (*Humanae Vitae*, 12).

Why does this matter? It matters that a human being is not reduced to a likeness to a robot. Or that an act which is intrinsic to

---

[16] Paragraph 48: https://www.vatican.va/archive/hist_councils/ii_vatican_council/documents/vat-ii_const_19651207_gaudium-et-spes_en.html.

husband and wife is not reduced to an account of its so-called similarities and differences to that of animals because, in the very nature of rendering it thus, there is a soul-less understanding of what God intended for marriage. And it is this soul-less understanding of matrimony which colours, in a colourless way, the possibility of a child being aborted or a thawed, frozen embryo being given a welcoming place by the heart of a married woman willing to adopt a child. In other words, the true "likeness" to human marriage is not that of robots, animals and the general laws of nature, but the mystery of *The Blessed Trinity*. As St. John Paul II said: 'Before creating man, the Creator withdraws as it were into himself, in order to seek the pattern and inspiration in the mystery of his Being, which is already here disclosed as the divine "We"' (*Letter to Families*, 6). And, reflecting or indeed even "imitating" the divine "We", the being of love, 'in its fullness, is inter-personal and not individual [...]. It is reciprocity which, in love, decides the birth of this 'we'. It proves that love has matured, has become something between persons, has created a community'[17]. Conversely, marriage is also a prophecy of the mystery of God: 'Monogamy is not archaism, but prophecy: it reveals that human love, lived in its fullness, somehow anticipates the very mystery of God'[18].

Thus, to draw on this emphasis on the new anthropology which reached a watershed, as it were, with St. John Paul II, we arrive at the theology of the person-as-gift. Thus the person-as-gift is the true heart of God's view of the person, being in Himself an everlasting

[17] *Una Caro*, paragraph 118, citing Karol Wojtyla, *Love and Responsibility* , trans. edited by A. Milanoli, Marietti, Genoa–Milan 1980, 61-62.

[18] *Una Caro*, paragraph 107.

and eternal gift of one divine Person being ever a "Gift" for the "Other", in the mystery of the Blessed Trinity: the three Persons in One God[19]. Thus the personalization of the law of gift, in referring to the reciprocal gift of husband and wife, reveals more clearly the heart of love, leading us to see that law is an expression, not just of love but of the mystery of divine love being embodied in the 'one flesh' covenant of Christian marriage. This is the true "hearth of love" to which conversion calls us and in which we are to welcome the gift of the child.

There is, too, the new context of the renewed emphasis on the mercy of God: on the 'ample arms' of Jesus Christ for all – particularly the unbaptized infant

---

[19] Cf. Karol Wojtyla, both in *Love and Responsibility* and *Person and Act,* was about developing the subjectivity of the subject. Then this work was strikingly enriched by Scripture and a more personalist understanding of human being in *Familiaris Consortio* and, indeed, his whole theology of the body catecheses. Therefore, cf. Etheredge, Chapter Five: "The Human Person in the Light of the Anthropology of St. John Paul II: Who is man? The Answer of Reason: the Rational Nature of the Person; and the Answer of Faith: Man, Male and Female in the Light of the Word of God and the Blessed Trinity. Finally, Man, Male and Female, Sin Together, Are Redeemed Together and Are "Taken Up" in Christ and His Church Together" in *Volume II: Faith and Reason in Dialogue*, of a trilogy on Truth and truth: https://www.cambridgescholars.com/product/978-1-4438-8911-7; and cf. Etheredge on the development of St. John Paul II's thought in *Scripture: A Unique Word*, Chapter One: "From Person to Person: A Reflection on the Work of Pope John Paul II and the Possibility of Interpersonal Communication": https://www.cambridgescholars.com/product/978-1-4438-6044-4.

"But infinite Goodness hath such ample arms
That it receives whatever turns to it"
(Dante, Purg. III, 118-123)' ([20]5).

While the problem of translation abounds, Dante is clearly referring to the Lord "who" receives whoever 'turns' to Him! But the Lord does not just let whoever 'turns' to him but, like the father in the parable of the prodigal son (cf. Lk 15: 20) – embraces all who turn to Him in His 'ample arms'![21]

We find, then, a new emphasis in the mentality and the documents of the Catholic Church on the mercy of God as a remedy for what have been, hitherto, challenges to her pastoral care and understanding. And, therefore, let us consider the new context for our investigation into God's love for the unborn, beginning with a wider consideration of how the mercy of God is showing itself in other areas of life. Let us remember, too, that this book is being written in the context of the *Year of Hope*, a hope that does not disappoint (Rom 5: 5), promulgated by the late Pope Francis and redolent with the expression of the abundant mercies of God shown in the many ways that we can receive the generosity of God's forgiveness, not just

---

[20] Pope Francis' Letter on Dante, *Cantor Lucis Aeternae*, paragraph 5: https://press.vatican.va/content/salastampa/it/bollettino/pubblico/2021/03/25 /0181/00393.html#ing.

[21] From Etheredge, *Time! Ladies and Gentlemen: A Few Thoughts on Last Things–In response to James Sale's Trilogy: HellWard, StairWell, and DoorWay*, forthcoming from En Route Books and Media, 2025-2026: https://enroutebooksandmedia.com/time/.

of our sins but also of the debt that we, or another, incurred through our imperfect contrition and which, without an indulgence, remains to be paid in full in purgatory. An indulgence is an expression of the treasury of forgiveness that Christ entrusted to His Church; and, in an indulgence, it is dispensed generously if certain conditions are met: sacramental confession; receiving the Eucharist; praying for the pope; and in a designated place. Thus, in the context of *a hope that does not disappoint*, we consider three ways that the Church is showing a renewed emphasis on the mercy of God: a better pastoral understanding of suicide; the blessing of individuals in irregular unions; and, particularly relevant to this book, the entrusting of unbaptized infants to the mercy of God.

Even the vexed question of how to respond, sensitively, both to the person who commits suicide and to the needs of his or her surviving family members, has been re-interpreted through the mercy of God. As it says in the new *Catechism of the Catholic Church*, not forgetting the gravity of the act of ending the life that 'God has entrusted to us' (CCC, 2280), the degree and subtlety of personal responsibility (CCC, 2282), or the scandal of encouraging others (CCC, 2282), the Church goes on to say: 'We should not despair of the eternal salvation of persons who have taken their own lives. By ways known to him alone, God can provide the opportunity for salutary repentance. The Church prays for … [people] who have taken their own lives' (CCC, 2283). And, speaking from personal experience, I know how sudden and without forethought the intention to kill oneself can arise, having tried to do it when I was about fourteen years old; and, owing to a particular grace of "seeing" the possibility of being judged by Christ and His apostles, I started drinking water and lived to tell of the mercy and glory of God[22]. In another case, a young woman committed suicide – but there was such an outpouring of prayer for her that I cannot but believe that it was inspired by God and that, even though she died, that this outpouring of prayer has contributed to her being found in the merciful arms of our Almighty God and *Father*.

In another instance, take the late Pope Francis' document on blessings, which has been widely misunderstood, emphasizing the need to bless a particular person without condoning an action or way of life that he or she may have chosen or be immersed in. Thus the document on Blessings states that those who, relying on the mercy of God, 'constantly seek a blessing from Mother Church': 'It is precisely in this context that one can understand the possibility of blessing couples in irregular situations and same-sex couples without officially validating their status or changing in any way the Church's perennial teaching on marriage'[23]. Thus we see an emphasis on the merciful action of Christ being expressed in the palpable mercy of the Church's blessing of particular people. Indeed, it is as if Christ the Good Shepherd is emerging more clearly in the life of the Church, emphasizing both compassion for the sinner and for

---

[22] Etheredge, *The Prayerful Kiss*: https://enroutebooksandmedia.com/theprayerfulkiss/.

[23] I have reordered the two statements from the document authorised by Pope Francis, "DICASTERY FOR THE DOCTRINE OF THE FAITH, Declaration *Fiducia Supplicans*, On the Pastoral Meaning of Blessings: https://www.vatican.va/roman_curia/congregations/cfaith/documents/rc_ddf_doc_20231218_fiducia-supplicans_en.html.

those close to the person who has either died or is in need of being helped to draw closer to the redeeming power of the love of God.

And so it is with the new mentality of the Church which now embraces the child who, through no fault of his or her own, dies without baptism. Thus the *Catechism of the Catholic Church* says: 'As regards children who have died without Baptism, the Church can only entrust them to the mercy of God, as she does in her funeral rites for them' (CCC, 1261). Naturally, then, this does not dispense from the obligation to baptize, whether in the ordinary or the extraordinary situations of life. But it does signify a change as the doctrine of limbo has been dropped from the teaching of the Church. Limbo expressed the idea that if a child died, unbaptised, and had not committed a personal sin, that it would be unjust for them to go to hell – even if they were still under the condemnation of God due to original sin. Limbo, then, was understood to be a natural state of happiness which was neither heaven nor hell. Now, however, there is an emphasis on the mercy of God and the possibility that God, not being bound by his sacraments, can act to save us through the merits of Jesus Christ, His only begotten Son in a way known only to Him[24].

We see, then, in the case of those who commit suicide. those who are in irregular unions but who can receive an individual blessing, and those who die as unbaptized infants, that there is this emphasis on the merciful action of God such that: 'By ways known to him alone, God can provide the opportunity for salutary repentance'

---

[24] *The Hope of Salvation for Infants who Die without Being Baptised.* https://www.vatican.va/roman_curia/congregations/cfaith/cti_documents/rc_con_cfaith_doc_20070419_un-baptised-infants_en.html#_ftnref130

(CCC, 2283). Thus, when considering the vast numbers of the unborn who, through no personal fault of their own, are destroyed, discarded, experimented upon or frozen, and who often die while being thawed, that there is a renewed, rescuing mentality, expressed in the Church, who seeks the salvation of all![25] Moreover, there is even a ministry for the dignified burial of children who, for one reason or another, are the "fallout" of the production process of *in-vitro* fertilization. This ministry is pioneered by Laura Elm, who founded *Sacred Heart Guardians and Shelter* (SHG) in 2017. Laura, it seems, while working 'on the business side of the infertility business' fell pregnant with her first child, who 'was born on March 25th, the Feast of the Annunciation'! This discovery of the discrepancy between Laura's sudden experience of motherhood and her work involving the 'waste' of many children led to this new ministry:

> [Laura] 'reaches out to doctors and embryologists at fertility centers and offers to provide burial for the embryos who die while in their care'[26].

Thus in a somewhat discreet but real way, Laura sows the seeds of recognizing the humanity of the human, embryonic child.

---

[25] *The Hope of Salvation for Infants who Die without Being Baptised*, paragraph 5, citing 1 Tim. 2: 4.

[26] You can read the full account of Laura's experience in her "Foreword to Chapter Three: Mary's Living Hope", in Etheredge, *Mary and Bioethics: An Exploration*, pp. 105-110: https://enroutebooksandmedia.com/maryandbioethics/

Truth and Legislation

‘The nine-judge court said in the 8-1 ruling that the [Alabama] state's "Wrongful Death of a Minor Act" is "sweeping and unqualified," and that its provisions extend to children "regardless of their location"’[27]. In other words, this ‘sweeping and unqualified’ defence of children, ‘regardless of their location’ is about as perfect as a law can be to legislate for a child’s protection; as the law is, as it were, directly ordered to the defence of the child and neither swerves one way nor another in mitigating that protection. However, in the aftermath of this decision, bills have been drafted to effectively exclude human embryos that ‘exist outside of a human uterus’ from the protection of the law[28].

And, as can be seen from the controversies that erupted in American courts, defending the human embryo in or out of the womb has to recognize that there is an enormous emotional and financial investment in the in-*vitro industry.* Therefore it is necessary to look at legislation that has been drafted elsewhere and succeeds, at least, to limit the obvious harm to embryonic children. And, indeed, the law in Italy that regulates the in-*vitro* fertilization industry, has endeavoured to safeguard the life of the child fertilized in a glass

---

[27] Daniel Payne, February 20th, 2024, “Alabama Supreme Court rules that frozen embryos are children under state law”: https://www.catholicnewsagency.com/news/256891/alabama-supreme-court-rules-that-frozen-embryos-are-children-under-state-law

[28] Etheredge and Elizabeth Rex, *The Zygote of Christ and the Mystery of Man*: Rex, Essay V, pp.117-135, p. 126: https://enroutebooksandmedia.com/zygoteofchrist/

dish. However, although some of its principles have been blunted, it still stands as an exemplary defence of the embryonic child:

> 'A maximum of three embryos could be fertilized and all had to be transferred to the mother's womb.
>
> - No preimplantation genetic testing (PGT) or "selection" of embryos was permitted.
> - Embryo freezing was prohibited, except in temporary emergencies.
> - No donor sperm, donor eggs, or any form of surrogacy could be used.
> - Informed written consent - prior to the fertilization process — required that all the embryos created would be maternally received as children, not discarded or frozen as "spares"'[29].

In another part of the world, this time in Puerto Rico, there is an interesting development of the concept of the 'natural person': that 'The measure [Senate Bill 504[30]] revises key provisions of the Civil Code to state explicitly that every human being is considered a natural person, a definition that now includes the human being

---

[29] Elizabeth Bothamley Rex, October 22nd, 2025, "Italy's Embryo Protection Law at 20: A Legacy of John Paul II and Benedict XVI": https://www.ncregister.com/commentaries/two-popes-italy-ivf-law.

[30] https://sutra.oslpr.org/medidas/155081.

conceived in the mother's womb at any stage of pregnancy'[31]. On the one hand this is a wonderful development, expressing the truth that the nature of human procreation or an outcome of spousal union, like twins, is the 'natural person': the person conceived through the natural acts of husband and wife. The question, however, can be asked, can this provision apply to the beginning of any human being, however he or she has begun, given that the human sperm and egg-cell are intrinsically orientated to the beginning of human personhood? In other words, to be a 'natural person' it is both necessary and sufficient to be conceived of what transmits and transforms the biological antecedents into a human person, namely a human sperm and a human egg-cell or their human equivalent. Unfortunately, it is necessary to add 'or their human equivalent' as, contrary to good science and the love of human life, there are experimental acts that endanger the integrity of human personhood.

In a remarkably prescient echo of this phrase, the 'natural person', there is what St. Athanasius said of the conception of Jesus Christ:

> 'Gabriel announced the good news to Mary with all clarity: he did not say simply: 'what is born in you', in case it might be thought that the body had been introduced into her from outside; he said: 'what is born of you', so that it would be

---

[31] ENRIQUE VILLEGAS, December 30th, 2025, "Puerto Rico Redraws the Legal Map of Prenatal Life: every human being is considered a natural person": https://zenit.org/2025/12/30/puerto-rico-redraws-the-legal-map-of-prenatal-life-every-human-being-is-considered-a-natural-person/.

accepted that what she gave birth to, came from her in the natural way'[32].

In other words, notwithstanding the mystery of Christ's conception not requiring a human father, but His conception being brought about by God (Lk 1: 35), the Author of human life, we see this emphasis on what is human, being from what is human.

In a word we need what the Constitution of the United States endeavoured to do, to protect the right to life[33], which is what recent documents continue to affirm, but which have not been clearly recognized, namely, that a human being is a human person from conception[34]. For the human embryonic child literally embodies the right to life. However, by contrast, the recent American endorsement of *in-vitro* fertilization is so indiscriminate as to ignore the very truth it has otherwise striven to express, namely, that each one of us is a person from conception and that, therefore, a procedure which extinguishes human beings, sends them for experimentation, or

---

[32] St. Athanasius: Letter to Epictetus, 5-9, on page 253 of *The Office of Readings* for the "Solemnity of Mary, Mother of God, January 1st, in *The Divine Office*, London: Collins, 1974 etc.

[33] Cf. Etheredge, *Unfolding a Post-Roe World*, wherein there is a discussion on the original meaning of the 14th Amendment, particularly parts IV-V of the book: https://enroutebooksandmedia.com/post-roe/.

[34] January 20th, Executive Order", "Presidential Actions: DEFENDING WOMEN FROM GENDER IDEOLOGY EXTREMISM AND RESTORING BIOLOGICAL TRUTH TO THE FEDERAL GOVERNMENT": https://www.whitehouse.gov/presidential-actions/2025/01/defending-women-from-gender-ideology-extremism-and-restoring-biological-truth-to-the-federal-government/.

freezes them, is actually anti-life – although it gives rise to children who are transferred to a woman![35] In other words, we need a new, international recognition of the right to life which is inalienably integral to each person who comes into existence.

Having set the scene, somewhat, for the legal defence of an embryonic child, we need to examine more exactly what can be known about human conception. As we will see, a closer examination of human conception vindicates the legal protection of the embryonic child by demonstrating that if a child comes to exist at conception then that child, automatically and unconditionally, has a right to be protected by law. But beyond what a public law does or does not state, is the truth of the natural law that requires us to "do good and to avoid harm"; and, as such, this is the measure of all laws pertaining to the defence of human life. Does the law, as regards life, ensure the good and protect from harm?

But, along with many other voices, we must keep in mind the gravity of the implications of the destruction of innocent human life; as Pope Leo XIVth reminds us:

---

[35] "Fact Sheet: President Donald J. Trump Expands Access to In Vitro Fertilization (IVF)", February 18th, 2025: https://www.whitehouse.gov/fact-sheets/2025/02/fact-sheet-president-donald-j-trump-expands-access-to-in-vitro-fertilization-ivf/; indeed, this contradiction seems to become clearer and clearer, cf. Doug Mainwaring, January 7th, "Pro-life leaders slam Trump for urging GOP lawmakers to be 'flexible' on Hyde Amendment": https://www.lifesitenews.com/news/pro-life-leaders-slam-trump-for-urging-gop-lawmakers-to-be-flexible-on-hyde-amendment/?utm_source=daily-usa-2026-01-08&utm_medium=email.

‘Mother Teresa of Calcutta, saint of the least and Nobel Peace Prize winner, stated in this regard that "the greatest destroyer of peace is abortion" (see Address at the National Prayer Breakfast, February 3, 1994). Her voice remains prophetic: no policy can serve people if it excludes from life those who are about to come into the world, if it does not assist those in material and spiritual need’[36].

[36] January 31st, 2026, “Pope Leo XIV Tells Young Politicians at Ḥackathon " there will be no peace without ending the war humanity wages on itself": https://www.catholicnewsworld.com/2026/01/pope-leo-xiv-tells-young-politicians-at.html.

Chapter Two

An Objective Account of Nascent Personhood

A closer examination of human conception, however, is also necessary because this event founds what can be understood about the inseparable relationship between the salvation of all and the salvation of embryonic children. Although while actual twinning, where one child takes a start from the fertilization of the first child, is another moment of beginning, this is not discussed beyond recognizing that, as soon as a child exists, so he or she exists in union with the *Incarnation* of Christ. So, focusing as this book does on the original act of fertilization, let us consider what has come to light.

Pope Francis says that the question of the personhood of the embryonic child is still being debated[1]. I therefore explain, briefly, why it is true that human personhood begins from the first instant of human fertilization. And, in so far as this question is open to debate, it follows that I am free to advocate a position not formally advocated by the Magisterium of the Catholic Church. However, let it be noted that St. John Paul II says something very closely aligned with what I will argue, albeit this is not the original language of his text but a translation:

---

[1] The Editors, November 28th, 2022, *America: The Jesuit Review*, "Exclusive: Pope Francis discusses Ukraine, U.S. bishops and more": https://www.americamagazine.org/faith/2022/11/28/pope-francis-interview-america-244225; Cf also St. John Paul II, *Evangelium Vitae*, paragraph 60.

> Modern genetic science '*has demonstrated that from the first instant there is established the programme of what this living being will be: a person, this individual person with his characteristic aspects already well determined. Right from fertilization the adventure of a human life begins, and each of its capacities requires time-a rather lengthy time-to find its place and to be in a position to act*'"[2].

Note that St. John Paul II says 'from the first instant there is established the programme of what this living being will be: a person'. But, if there seems ambiguity in the expression, 'what this living being will be', implying that this living being is not, or possibly not yet, a person from the first instant of fertilization, then this justifies my pursuit of the following question. When, exactly, does human personhood begin? On the other hand, 'person' may be understood as the goal of this process; and, in effect, the manifestation of the person cannot be the goal of a process if the person is not present from the beginning. So, in the end, there are two options: either there is an immediate, first instant of fertilization's beginning of human personhood; or, alternatively, personhood comes to exist at some later stage, whether at the unification of the nuclei of the sperm and egg, which are now not separate items anymore, or at some later stage of embryological development.

Why do I introduce the expression: the manifestation of the person present from conception? Because, as a chorus of voices maintains, the walled, human embryonic child, does not look like a

---

[2] *Evangelium Vitae*, 60: italics added.

person; and, therefore, human development is about manifesting, showing what is there from the beginning. For the outcome of the manifestation of the person is made possible because of the existence of the human person; and the existence of the human person, from the first instant of fertilization, is caused by a proportionate cause, namely, God. Thus, the biological gametes of sperm and egg-cell, while they are the constituents of both the human body and the dynamism of the new human life, they are inadequate in themselves to account for the personhood of the human child. This is because both egg-cell and sperm are expressions of the transmission of human, biological life, but they are not, in themselves, transmitting human personhood. Human personhood, arising out of the whole entity, has to have a different origin to that of the biological "parts", namely sperm and egg-cell; for a whole cannot arise out of a part: the life of a person cannot arise out of biological life. Personal life transcends biological life. Therefore, there has to be a cause proportionate to the existence of a person. A human person, then, is begotten through the action of God. Moreover, bearing in mind the sign-language of the body and the sacramental acts of God, it is clear that the living, outward sign of the formation of the embryonic wall, is a possible outward sign of the action of God which brings a human embryonic child to exist.

God acts to make Himself known. The moment most characteristic of an action of God is when there is an outward sign of an inward act, as with the words and actions which constitute the ordinary rite of Baptism. Thus there is the pouring of water on the head and the words which, together with the pouring of the water, draw us into the mystery of the Blessed Trinity: I Baptise you in the name of the Father and of the Son and of the Holy Spirit. Similarly, as the Creator is the author of both natural and supernatural acts it makes sense, therefore, that the first instant of fertilization, whereby the human embryo is "encased" in his or her embryonic skin, is the natural outward sign of the inward act of God which brings the whole person to exist, one in soul and body (cf. *Gaudium et Spes*, 14)[3]. And, in sum, just as each one of us is conceived in-relationship, so we are conceived for relationship: for a relationship to God and to each other.

I advocate this position, of the reality of human personhood from the first instant of fertilization – both because I argue it is true in itself and because it is completely relevant to this book on *The Salvation of Embryonic Person*. For, if it is true in itself, that the first instant of fertilization is the absolute beginning of human personhood, it makes sense of the fact that each human life has to have an actual beginning and not some arbitrary demarcation between one stage of human development and another. An absolute beginning of human personhood, however, does not mean tracing human personhood back beyond conception. While the nature of the egg-cell and the sperm are completely relevant to the conception of a human being, conception is precisely the coming together of egg-cell and

---

[3] While there are surveys on the meaning of conception, fertilization, pregnancy, they do not seem to me to be sufficiently concrete, in the light of modern embryology and theological reflection on the conception of Mary and Jesus Christ; such surveys can also have a deliberately polemical bias in favour of abortion and abortifacient contraceptives: Rachel Benson Gold, Guttmacher Institute, May 5th, 2005, "The Implications of Defining When a Woman Is Pregnant": https://www.guttmacher.org/gpr/2005/05/implications-defining-when-woman-pregnant.

sperm and not the fact that each of them exists independently of the other. The sperm, then, tiny and yet moving as it is, in comparison to the bulk of the inert egg-cell, is yet capable of initiating a transformation: that two independent entities are now one. No sooner does the sperm enter a pore of the egg-cell, than the egg-cell ceases to be what it is: inert; and, in ceasing to be inert, it encases the sperm head, setting up an ionic calcium reaction which traverses the whole entity and brings about, in a fraction of a second, the embryonic cell wall which spatio-temporally expresses the existence of what is now the human embryonic child.

The originating moment of the human embryo is no less relevant to the identity of Mary, who is conceived without original sin; and, even if she is conceived without original sin Mary, as a human child, is conceived as we are conceived. So this is a first step in a theological sequence. Moreover, it is argued elsewhere, that Mary cannot be conceived without original sin if she is not 'one in body and soul' (*Gaudium et Spes*, 14) from the first instant of fertilization'[4].

On the one hand, the *Immaculate Conception* of Mary, and the *Incarnation* of the Son of God, are unique acts of God. Mary, from the first instant of her existence, exists whole and entire and free of original sin. The Son of God, from the first instant of His *Incarnation*, exists whole and entire and free from original sin. And, as St. Gregory Nazianz says, in his *Letter of Gregory to Cledonius the Priest, Against Apollinaris* (101):

---

[4] Cf. Etheredge, Chapter Five of the following book: *Mary and Bioethics: An Exploration*: https://enroutebooksandmedia.com/maryandbioethics/.

'That which was not assumed has not been healed; but that which is united to God, the same is saved'[5].

In other words, not only is it true that Christ heals what He assumes, but in so doing He does it for all human beings – from the first moment of their existence. Thus, it is of universal salvific significance that Christ is conceived from the first instant of fertilization, however that fertilization took place in the mystery of the *Incarnation.* This will be further discussed in the course of this book. Thus Mary and Jesus echo in a new way the original creation of Adam and Eve.

On the other hand, in the case of every other human being, we are conceived in original sin which is 'transmitted by propagation to all mankind, that is, by the transmission of human nature deprived of original holiness and justice' (CCC, 404). But, nevertheless, we too exist whole and entire from the first instant of the creative act of God; however, in our case, we have received, too, the deficiency of 'original holiness and justice' and are, therefore, 'wounded' in our humanity (CCC, 405). Hence, generally, the need for each one of us to be baptized; and, therefore, to receive the free-gift of beginning to be reconciled to God. Nevertheless, it can be said that 'as human

---

[5] *The Faith of the Early Fathers*, vol 2, passages selected and translated by William Jurgens, (Collegeville, Minnesota: The Liturgical Press, 1979), p. 41.

conception entails an act of God it is almost intrinsically ordered to salvation history'[6].

Donum Vitae and the first instant of human conception: The Latin edition of Donum Vitae, as distinct from the English translation

What relationship, then, is there between this argument for the definition of human conception to be from the first instant of fertilization and another document, *Donum Vitae: The Gift of Life*. Therefore, in the following, brief discussion, there is an examination of an expression used in the English translation of the Latin, authoritative edition, of *Donum vitae*. We begin with the English translation as it uses an expression which is not in the Latin text; and, as such, could be said to give a misleading impression of what, then, the Latin text actually said.

> Concerning the meaning of human conception, the English translation says: '*Right from fertilization is begun the adventure of a human life, and each of its great capacities requires time ... to find its place and to be in a position to act". This teaching remains valid and is further confirmed, if confirmation were needed, by recent findings of human biological science which recognize that in the zygote* resulting from*

[6] Etheredge and Elizabeth Rex, *The Zygote of Christ and the Mystery of Man*: Etheredge, Essay III, p. 54: https://enroutebooksandmedia.com/zygoteofchrist/.

> *fertilization the biological identity of a new human individual is already constituted*"[7].

> Furthermore, in the English translation, there is a comment below, picking up on the word which is starred, namely, 'zygote*'. Thus the starred comment reads: '*The zygote is the cell produced when the nuclei of the two gametes have fused*'[8].

But first a note about the word and the entity: zygote is a Greek word which means: 'joined, yoked'[9]; and, therefore, zygote is appropriately used to refer to the coming together of sperm and egg-cell. However, the first problem is that the English translation of a Latin note says, as we have said, 'The zygote is the cell produced when the *nuclei of the two gametes have fused*'[10]. But even if a zygote is formed by the fusion of the two gametes, the sperm and the egg-cell, the unification of the nuclei is a process which had a prior beginning, namely, 'The male and female pronuclei [of the sperm and the egg] are converging, but the genetic material is not yet united'[11]. In other words, the first instant of fertilization is prior to the fusion of the

---

[7] *Donum Vitae*, English text: "1: WHAT RESPECT IS DUE TO THE HUMAN EMBRYO, TAKING INTO ACCOUNT HIS NATURE AND IDENTITY?": http://www.vatican.va/roman_curia/congregations/cfaith/documents/rc_con_cfaith_doc_19870222_respect-for-human-life_lt.html. Italics added.

[8] *Donum Vitae*, 1.

[9] "Zygote": https://en.wikipedia.org/wiki/Zygote.

[10] The italics are added for emphasis.

[11] "Zygote": https://en.wikipedia.org/wiki/Zygote. But further discussion of this is beyond the main point of what concerns me here.

nuclei and the formation of the zygote; and, even then, the zygote is yet another stage on what is now an uninterrupted path of development.

Thus the question of whether or not the 'nuclei of the two gametes have fused', in this account, raises the question of what exists at the first instant of fertilization which, uninterruptedly, will manifest the presence of a person. In other words, the first stage of a zygote entails 'converging … genetic material' and, subsequently, that the 'nuclei of the two gametes have fused'; but, in either case, these embryological steps are after the first instant of fertilization. It is the first instant of fertilization which starts the whole process of the manifestation of the person. If, then, these accounts assert that human personhood is after the first instant of fertilization, then it raises the "problem" of a "pre-embryo"; that is, a non-personal, human entity, that then becomes a human person.

In view, however, of the argument of Mary's *Immaculate Conception*, this makes her 'pre-embryo' contaminated by original sin and is, therefore, contrary to the dogma of her *Immaculate Conception*. Furthermore, if this were a natural stage of embryonic development then Christ would have had to have a stage of being a 'pre-embryo' and that, again, would mean that the flesh of Christ's pre-embryo would also be contaminated by original sin. In other words, both the dogma of *the Immaculate Conception* and the *Incarnation* of the Son of God imply and require a whole and holy humanity from the first instant of the action of God which brings each of them to exist. Thus, prompted by these concerns, it was necessary to examine the authoritative Latin text.

> The Latin edition of *Donum vitae* says, at footnote 27: 'Zygotum est cellula orta a fusione duorum gametum'[12]: the Zygote is the cell arising from a fusion of the two gametes[13].

In other words, when we compare the two notes, one in the English text and one in the Latin text, there are significant differences. The English translation of the Latin footnote uses the words, 'The zygote is the cell produced when *the nuclei of the two gametes have fused*'[14]. In other words, the English translation of footnote 27 introduces an expression which is not in the Latin, namely, 'the nuclei of the two gametes have fused'. The Latin simply says a zygote, 'arising from the fusion of the two gametes'; and, therefore, there is no mention of the fusion of the nuclei of the two gametes. This is crucial because if the Latin is the authoritative text, then the Church has prudently left open the possibility that the Latin can be understood to include the very first instant of fertilization – prior to the fusion of the nuclei of the gametes[15].

---

[12] *Donum Vitae*, Latin text: http://www.vatican.va/roman_curia/congregations/cfaith/documents/rc_con_cfaith_doc_19870222_respect-for-human-life_lt.html.

[13] This translation of the Latin is deemed 'correct' by the Rev. Dr. Richard Conrad, OP, of the Oxford Dominicans (email, 17th of September 2025); Fr. Richard is an Official Fellow, Lector, Member of the Aquinas Institute, Oxford.

[14] Italics added.

[15] For this research and a look at other language editions, such as the German, French and Italian, see Etheredge, *The Human Person: A Bioethical Word*, Chapter 7: Part V of Section (II): The "Revisability" of the Definition of the Term 'Zygote' in *Donum Vitae*: https://enroutebooksandmedia.com/bioethicalword/.

On the other hand, it is clear that any embryological stage of development has, as it were, a beginning and an end; however, the question of the beginning of human life transcends embryological "stages" of development. It would be like saying there is more than a developmental difference between a child, a youth, an adolescent, a young man and an adult; however, each is a passing phase, passing into the next stage of life but all are consecutive on the very beginning, without which there would be no continuation, no development – nothing.

Further support for this view comes from reflecting on the mystery of the *Immaculate Conception.* St. Thomas Aquinas says that grace requires the presence of the soul; and, therefore, if the soul of Mary is not present from the first instant of her conception, then grace is not present either. And if grace is not present there is the possibility of the transmission of original sin, through the transmission of life from Mary's parents to her. The transmission of original sin is the transmission of the deficit of original grace; and, therefore, a kind of contaminating tendency to sin. Thus the dogma of the *Immaculate Conception* implies Mary's conception from the first instant that the sperm locks into the egg-cell and life is transmitted, the fruit of which is a walled, human embryo. Therefore Mary's *Immaculate Conception* has to be from the very first instant of fertilization if she is to be free of original sin and completely and wholly graced[16]. Similarly, in the Incarnation of the Son of God, given that the "Person" of the Son of God is enfleshed, it follows that He is

[16] Therefore, see Etheredge, *Mary and Bioethics: An Exploration*, Chapter 5.

united with a whole human nature and not some pre-embryo part of it; for, again, the grace of union requires the presence of a true subject, namely the soul of Jesus Christ.

As regards the possible objection that the human embryo, from the first instant of fertilization, is undifferentiated, consider the following analogy. A deep-sea diver is wearing breathing equipment when he or she is underwater. We do not destroy the breathing equipment because the diver depends on it while underwater. Similarly, with respect to the nascent human person, there is an integral human presence, entailing both a human person and his or her life saving equipment. Similarly, then, we can recognize and respect the integral nature of the development of the human embryonic child and the placenta that makes it possible and which is discarded at birth. So, while the diver is submerged we would not cut off the diving equivalent: diver and equipment are one but can be differentiated. And, therefore, embryonic child and placental development are one until it is time for their separation.

In a word, Pope Francis has not said that a human being is not a person; and, if there is uncertainty, then it entails acting in accordance with the good of human life (cf. *Evangelium Vitae*, 60).

On the dignity of each one of us and its violations

> *'The human person has been endowed with a very exalted dignity, which is rooted in the intimate bond that unites him with his Creator: a reflection of God's own reality shines out in the human person, in every person, whatever the stage or*

condition of his life'[17]. Thus this applies, whether the child is conceived in the mother or is one of the "extrauterine unborn children". Indeed, more explicitly, 'According to Bishop Sgreccia, who heads the *Pontifical Academy For Life, the human embryo, even if it is not being nurtured in a maternal uterus, is a child*'[18].

Establishing that the first instant of fertilization is the very beginning of conception, which itself means a beginning, and that this is the moment which outwardly expresses God's creation of the whole human person, makes all of us equal, whether born, unborn, or a frozen embryonic child. And, indeed, while a child has the right to be conceived within the embrace of spousal love as this, ordinarily, enables the child to be carried in the woman who becomes a mother, no child is without the dignity of being a child of God, however he or she comes to exist. Nevertheless, there are actions which undermine the perception of the natural and inherent dignity of the child. On the one hand, there is a general tendency to promote and encourage the abortion of a child, which creates a fundamentally, and widely destructive "throw away"

---

[17] Address of His Holiness Benedict XVI to the Participants at the 12th General Assembly of the Pontifical Academy for Life and Congress on "The Human Embryo in the Pre-Implantation Phase,", 27th February, 2006: https://vatican.va/content/benedict-xvi/en/speeches/2006/february/documents/hf_ ben-xvi_spe_20060227_embrione-umano.html.

[18] February 27th, 2006, "Pope asserts the sacred and inviolable dignity of human life even before birth": https://archivioradiovaticana.va/storico/2006/02/27/pope_asserts_the_sacred_and_inviolable_dignity_of_human_life_even/in2-68208. Italics added.

mentality of one of the most precious gifts, namely life. What is more, abortion is promoted by those who have been given the great gift of life and who use it to destroy that same gift as it is given to others. On the other hand, 'The [*in-vitro* fertilization] IVF industry treats human beings like products and freezes or kills millions of children who are not selected for transfer to a womb or do not survive'[19]. But none of these actions diminish the dignity of each human person; even if there is both a tragic outcome for the child and an impact, however unobserved, on the conscience of the abortionist or IVF practitioner. As the Church says:

> 'the Church, … ardently urges that *respect for the dignity of the human person beyond all circumstances* be placed at the center of the commitment to the common good and at the center of every legal system'[20].
>
> And again, 'Indeed, respect for the dignity of each person is the indispensable basis for the existence of any society that

---

[19] "IVF Destroys Human Life, Bishops Urge Ethical Alternatives", paragraph 64: https://www.usccb.org/news/2025/ivf-destroys-human-life-bishops-urge-ethical-alternatives.

[20] "Bishop Chairmen Respond to Administration's Announcement of Expansion of Access to IVF and Fertility Treatments": https://usccb.org/news/2025/bishop-chairmen-respond-administrations-announcement-expansion-access-ivf-and-fertility.

claims to be founded on just law and not on the force of power'[21].

Which word, 'power', reminds us where this book began with Pope Leo referring to 'contact with those who have neither power nor greatness is a fundamental way of encountering the Lord of history'[22]. And who can be more without power than those upon whom death, experimentation, or freezing, is imposed?!

And who was one of those who had 'neither power nor greatness'? Each and every embryonic child was without power or greatness, barely conceived in the womb or fertilized in a glass dish, and vulnerable to any and every kind of abuse. Our original vulnerability, in other words, is what we not only have in common with each other but with the Lord, the Son of God, and His Virgin Mother Mary.

And lest we think that the Lord of history was any the less vulnerable to death than the rest of us let us both remember that He died a terrible and tragic death and that He barely avoided dying as an infant, shortly after birth, if it were not for Joseph receiving a message from an angel (cf. Mt. 2: 13-18). Maybe this generation needs to be visited by angels: messengers of hope!

---

[21] "Declaration of the Dicastery for the Doctrine of the Faith "Dignitas Infinita" on Human Dignity, 08.04.2024": https://press.vatican.va/content/salastampa/en/bollettino/pubblico/2024/04/08/240408c.html.

[22] *Dilexi Te*, paragraph 5.

Chapter Three

The Annunciation and the Eternal Significance of Embryonic Human Life

In this chapter there is first of all an exploration of how God acts: that there is a priority of the action of God which brings a child to exist – both in a relationship to his or her parents and to God. There is then a discussion of how the moment Mary conceives her Son, Jesus Christ, helps us to see that all children are conceived in relationship to the *Incarnation*; and, therefore, the children conceived *in-vitro*, in a glass dish, are also brought to exist in a relationship to God. Furthermore, then, there is a consideration of the possibility that the adoption of an orphaned human embryo by a married woman is, therefore, by extension, a saving act. Indeed, the more we reflect on the multiple acts of God which, as it were, come to meet us, the more possible it seems that God will not neglect the least, the most powerless, among us.

> "'"And behold[1], you will conceive in your womb and bear a son, and you shall call his name Jesus. He will be great, and will be called the Son of the Most High; and the Lord God will give to him the throne of his father David, and he will reign over the house of Jacob for ever; and of his kingdom there will be no end." And Mary said to the angel, ... ["But

[1] I prefer this translation, 'And Behold', as it seems more becoming than a bare 'Look!', cf. *The Holy Bible*, Catholic RSV translation, San Francisco: Ignatius Press, 1966.

> how can this come about, since I have no knowledge of man?"[2]] And the angel said to her, "The Holy Spirit will come upon you, and the power of the Most High will overshadow you; therefore the child to be born will be called holy, the Son of God ..."' (Lk. 1: 31-35)[3].

As we move closer to the inseparable interrelationship between the salvation of each one of us, embryonically, and the salvation of all, we come to the *Annunciation* which is, as it were, the heart of our subject where the Lord Jesus Christ first encounters each one of us. For, if it is true that Christ has united himself in a certain way with each one of us (cf. *Gaudium et Spes*, 22), then it follows that all of us have been united, in a certain way, to Christ. Even Adam and Eve, in their originally graced state of integral health and holiness, would only be given the grace of their original integrity in view of the death and resurrection of Jesus Christ Himself. Even if, however, it is difficult for us to understand the precise relationship between the *Incarnation* of the Son of God and the grace given to Adam and Eve and, subsequently, to Mary, albeit including every act of salvation in between, nevertheless it is in view of Jesus Christ's death, resurrection and ascension that any of us received what is given by Christ: 'grace upon grace' (Jn. 1: 16). This could be expanded indefinitely in that the whole history of salvation, while it has

---

[2] But I prefer *The New Jerusalem Bible's* translation of 'no knowledge of man' of the Greek phrase, 'ou ginwoskw', as it is closer to the Greek and echoes Eve in Genesis, to be discussed later; *NJB*, London: Darton, Longman and Todd, 1985.

[3] *The Holy Bible*, Catholic RSV translation, 1966.

"highlights", such as Elizabeth and John the Baptist's response to Mary, and the Lord she carries, is a history that expresses God's willingness to step over, as it were, our deficit of original grace and grant us both existence and, in many cases, to be a witness to His actions.

> 'Let us note, too, what St. John Chrysostom says: 'And that first tree sent men to Hades; this second one ['the tree of the cross'] calls back even those who had already gone down there. Again, the former tree concealed man already despoiled and stripped; the second tree shows a naked victor on high for all to see. And that earlier death [of Adam and Eve] condemned those who were born after it; this second death [of Christ] gives life again to those who were born before it'[4].

On the one hand, what St. John Chrysostom says endorses the reach of Christ's death and resurrection in that He 'gives life again to those who were born before it'; and, therefore, St. John encourages us to think trans-temporally: across the whole of time. On the other hand, it not only makes it explicit but also leaves open the very possibility that the original grace in which Adam and Eve were created was of Christ's death and resurrection.

It is this relationship, then, between Christ and "all of us" that both answers the question of Mary's *Immaculate Conception* and the

---

[4] A Reading from the homilies of St. John Chrysostom: On the burial place and the Cross 2, from the Saturday Memoria of Blessed Virgin Mary, p. 256* of The Divine Office, Volume I: Advent, Christmastide and Weeks 1-9 of The Year, London: Collins, 1974.

possibility of the salvation of all, especially of those who have died without baptism. Can we therefore inquire, as it were, through the *Incarnation*, as to this way of salvation that is known only to God? Our starting point is that all of us, through His *Incarnation*, have been united, in a certain way, to Christ (*Gaudium et Spes*, 22).

But, before exploring the *Annunciation* further, we begin by clarifying how it is that God the Father's paternity does not contradict that of Joseph's, in the *Incarnation*.

Two different kinds of paternity: God the Father's and Joseph's

> 'Mary consented to conceiving Christ without the involvement of her spouse, Joseph. Though they were not yet living together, under Jewish law, Mary and Joseph were already considered husband and wife'[5].

A question for us to keep in our hearts, then, is that it is God who reconciles His paternity with the vocation to fatherhood of St. Joseph, the spouse of Mary, in such a way that the divine Fatherhood of God does not contradict the spouses 'exclusive right to become father and mother solely through each other'. On the one hand, we know from Scripture that both Mary and Joseph remained virgins in the human sense of not knowing each other. On the other hand, God does not come between Mary and Joseph's 'exclusive right to

---

[5] Whitney Anderson's, Fall 2019, University of St. Thomas, Master's Thesis, "Justice for the Orphan: the Ethics of "Embryo Adoption"", p. 35: https://ir.stthomas.edu/cgi/viewcontent.cgi?article=1022&context=sod_mat.

become father and mother through each other'. The action of the Holy Spirit 'overshadowing' Mary, while effective in bringing about the *Incarnation* of the Son of God, is non-carnal and therefore not of the flesh. Whatever the precise nature of the mysterious act of the *Incarnation*, then, it is not to be confused with the ordinary human conception of a child which, in terms of the human, spousal act, draws on the reciprocal gift of husband and wife.

At the same time, when we are talking about the conception of Mary, from the action of God, Joachim, and Anne, we are talking about the beginning of the whole person: a beginning characteristic of both Mary and indeed of all of us. In the case of Jesus Christ, we are talking about the complete union of the both the Person of the Eternal Word, the Son of God, and the whole of human nature, one in body and soul (*Gaudium et Spes*, 14). This is because the eternal nature of the Son of God, who becomes *Incarnate*, enfleshed in the flesh of Mary's egg-cell or ovum, has no beginning as the eternal Son – but has a beginning as an "enfleshed" man. And, as we shall see subsequently, there is an important priority of the action of God which nevertheless comes through reflecting on the *Incarnation* of the Son of God.

On the one hand, then, this emphasis on the *Incarnation* of Christ, in the context of Mary's virginal marriage to Joseph, is not a condemnation of the 'one flesh' (Gn. 2: 24)[6] union of husband and wife. For, in its own way, the ordinary means of the transmission of

---

[6] This expression is used in the *Navarre Bible, Pentateuch*, Editorial Committee, Jose Maria Casciaro *et al*, for the English edition, James Gavigan *et al*, Dublin: Four Courts Press, 1999, p. 51: Genesis 2: 24, whereas the *Jerusalem Bible* says, 'one body' Genesis 2: 24.

life is also a sign of the mystery of the Blessed Trinity in which each Person lives, as it were, the reality of *life from Life.* Thus spousal union is itself an expression of the language of the body which transcends the natural order, implying that the person conceived is conceived from the reciprocal gift of husband and wife and is, therefore, a gift in his or her own right: flesh from flesh and life from life[7]. Indeed, the Church says of the human person, not just of the human individual but of the very nature of man as a being-in-relationship that:

> 'the Lord Jesus, when He prayed to the Father, "that all may be one. . . as we are one" (John 17:21-22) opened up vistas closed to human reason, for He implied a certain likeness between the union of the divine Persons, and the unity of God's sons in truth and charity. This likeness reveals that man, who is the only creature on earth which God willed for itself, cannot fully find himself except through a sincere gift of himself[8].

Let us note, too, that just as 'man ... is the only creature on earth which God willed for himself', so it is clear that each and every human person is directly willed by God. And, if there is need of an explanation as to how God wills the existence of each one of us, it is that the action of God is necessary to the existence of each one of us. As the *Catechism of the Catholic Church* says:

---

[7] Cf. Etheredge, *Scripture: A Unique Word*, Chapter 12: "Life "from" Life: A Reflection on the Moment a Person Comes to Exist": https://www.cambridgescholars.com/product/978-1-4438-6044-4.

[8] *Gaudium et Spes*, 24.

'every spiritual soul is created immediately by God – it is not 'produced' by the parents' (CCC, 366).

Thus there is no such thing as an unwanted child; for just as it is "natural" to God to will each of us into existence, so it is natural for us to welcome the child whom God wants to exist. The problem is that we do not act according to our nature and thus many, many children are in fact rejected, either in utero or in the course of the process of *in-vitro* fertilization, as some are transferred to a woman or the child's mother and some are rejected.

Marital union is, then, ordinarily, an enfleshed expression of the mystery of the Blessed Trinity in which each person is from the Other; and, while a child is from the flesh of his or her parents, so he or she is also from the "Other" who is God. And it is God who brings him or her to exist as a whole human person from the first instant of fertilization. However, as it has been said, embryological and human development is a process which culminates in the manifestation of the person present from conception; and, therefore, it is no contradiction to this being so that the human, embryonic child, no more resembles the adult human being than a germinating seed resembles the fully grown plant or tree. This is true of both Jesus Christ in His humanity, Mary, and all whom the Lord knew, saying "Before I formed you in the womb I knew you' (Jer. 1: 5). Also, 'Thy eyes beheld my unformed substance … [and] the days that were formed for me, when as yet there was none of them' (Ps. 139: 16, Catholic RSV). Thus, together, these verses fructuate, are fruitful in their interaction, suggesting that God knew the person whom He had brought to exist even before any development took place; and, therefore, from the very first instant of His creation of them. And what moment is there before any development takes place when, 'as yet there were none of them'? The first instant of our existence from the creative act of God.

On the other hand, what is unique about the relationship of Mary to her son, Jesus Christ, who is the Son of God, is that human flesh is so acceptable to God that He took it upon Himself to become, literally, a man born of woman. This endows both man and woman with an unmistakable and irrevocable dignity expressing, in a wonderfully literal way, that man and woman are not accidents of nature but the deliberate gift of God. Moreover, both in the *Incarnation* of the Son of God and the *Immaculate Conception* of Mary, we see that what is pre-eminent in the identity of the human person is the action of God. Let us, therefore, see, that just as the identity of Jesus Christ and Mary are from the action of God, so is the identity of each of one of us from the loving action of God. And, even if there is a problem in the expression of our parents' love for us, let us remember that God loves us unconditionally as Father and the Church as Mother! And that God wanted us from all eternity!

This priority of the action of God in bringing the human person to exist is both ancient and new. The priority of the action of God in the creation of the human person is pre-eminently expressed in the creation of Adam and Eve. In a particularly clear way, however, the priority of the action of God is expressed in the consciousness of Eve, who says: 'I have gotten a man with the help of God' (Gn. 4: 1). Clearly, Eve understood the action of her husband, Adam, as it says: 'Now Adam knew Eve his wife, and she conceived and bore Cain, saying, "I have gotten a man with the help of the Lord' (Catholic

RSV., Gn. 4: 1). In other words, it is possible to argue that Eve "sees" that both being "known" by Adam and, also, 'the help of the Lord' are, in fact, what brings about the conception of her child, Cain; however, it is also clear to her that there is a priority to the action of God in the immediacy of the cause of Cain's existence: that Cain exists precisely because Eve received 'the help of the Lord'.

> *How does the action of God in the Annunciation help us to understand God's relationship to the human embryo and to the possibility of the adoption of a human embryo being, literally, a saving act: an act of salvation?*

> 'God's mercy accompanies us daily. To be able to perceive his mercy it suffices to have a heart that is alert. We are excessively inclined to notice only the daily effort that has been imposed upon us as children of Adam[9].

In the first place, we must consider God's relationship to the human embryo; and, indeed, this is expressed very concretely in the life of Jesus Christ as He "entered" into human, embryonic life, in the very act of becoming *incarnate*. Thus, prior to discussing embryo adoption, it is important to recognize that there is the event of the *Incarnation*: that embryo adoption takes place in the context of salvation history: in the context that Christ, in a certain way, united

[9] Pope Benedict XVI, "MASS OF THANKSGIVING IN REMEMBRANCE OF THE POPE'S 80th BIRTHDAY: HOMILY OF HIS HOLINESS BENEDICT XVI": https://vatican.va/content/benedict-xvi/en/homilies/2007/documents/hf_ben-xvi_hom_20070415_80-genetliaco.html.

himself to every man (*Gaudium et Spes*, 22). Although it is too true that we live in a fallen world, it is also true that we live in a fallen world in the perspective of the resurrection: that our fallen world is embraced by an ongoing work of redemption[10] which receives both a definitive completion and a new beginning in the coming of Christ[11].

And, in drawing upon the Annunciation for inspiration[12], we do so in view of a whole history of God's action which makes the infertile, fertile[13]; and, even going beyond this, we draw upon the mystery that the Son of God takes human flesh and enters, wholly, into the embryonic personhood of man – albeit without a human father. Thus, although the Personhood of the Son of God is from all eternity, the humanity of Christ is from the first instant of the *Incarnation*. And, therefore, from the first instant of the *Incarnation* Christ

---

[10] Cf. Pope Francis, Opening paragraph of *Humana Communitas*.

[11] Cf. *Dei Verbum*, 4: https://www.vatican.va/archive/hist_councils/ii_vatican_council/documents/vat-ii_const_19651118_dei-verbum_en.html.

[12] Cf. Etheredge, "Embryo Adoption and Reciprocal Human Relationships", *To the Editor, The National Catholic Bioethics Quarterly* of America, Summer 2017; cf. also Kent Lasnoski, who is referred to in "An Overview of the Embryo Adoption Debate", Teofilo Giovan S. Pugeda, III, *PHILIPPINIANA SACRA*, Vol. LVII, No. 173 (May-August, 2022) pp. 187-214, but here pp. 208-209. https://philsacra.ust.edu.ph/admin/downloadarticle?id=6EEA626C18846E359AA63E549A534ECE.

[13] Cf. Etheredge, *Mary and Bioethics: An Exploration*, St. Louis, MO: En Route Books and Media, 2020, "Chapter Seven: Love, Scripture, Suffering and Bioethical Questions", pp. 229-254; but cf. also Cardinal Ratzinger, now Pope Emeritus XVI, *Daughter Zion,* translated by J. M. McDermott, SJ, San Francisco: Ignatius Press, 1983).

receives the gift of human nature, ensouled from the first instant of coming to exist. As 'St. Cyril of Alexandria and the third ecumenical council ... confessed "... the Word, uniting to himself in his person the flesh animated by a rational soul, became man"' (CCC, 466). For, as was recognized, the Son of God did not enter into union with an animal or a plant – but with what is characteristic of human nature: entailing a beginning to being wholly man, one in body and soul (cf. *Gaudium et Spes*, 14). As St. Gregory Nazianz says, in his *Letter of Gregory to Cledonius the Priest, Against Apollinaris* (101): 'That which was not assumed has not been healed; but that which is united to God, the same is saved'[14]. So, just as the Son of God 'became flesh' (Jn. 1: 14), so the flesh He became was constituted at the instant of His becoming man. In other words, there was no point at which the humanity of Christ existed without Him being wholly human; and, at the same time, there was no point at which the Son of God was not wholly united to His whole humanity. Thus, in effect, we end up with the argument that the Son of God's *Incarnation* was from the first instant that His ensouled flesh came into existence.

Therefore, entering wholly into human development, Christ redeemed us from the first instant of conception: from the first instant of His own existence as a human being which, embracing us all, constitutes the basis of His communion with all who come to exist[15]. At

---

[14] *The Faith of the Early Fathers*, vol 2, passages selected and translated by William Jurgens, (Collegeville, Minnesota: The Liturgical Press, 1979), p. 41.

[15] Cf. Etheredge, *Conception: An Icon of the Beginning*, St. Louis, MO: En Route Books and Media, pp. 220-221, drawing on Fr. John Saward's *Redeemer in the Womb*, Ignatius Press: San Francisco, 1993, pp. 8-13.

the same time, Christ embraced His relationship to His mother and, by implication, opens His heart to the full family reality through which each person comes to exist.

There is, then, a threefold logic to what follows. The Son of God has entered, wholly, into human nature, beginning from the first instant of being conceived by the Holy Spirit; and, therefore, everyone, from conception onwards, is embraced by His act of redeeming love. As the Greek Expositor says: '*For the Divine Word came to purify man's nature and birth, and the first elements of our generation. And so without sin and human seed, passing through every stage as we do, He is conceived in the flesh, and carried in the womb for the space of nine months*'[16]. And so, returning to the moment of the *Incarnation*, we read that it is through the *Incarnation* of the Son of God that he, in a certain way, united himself with each one of us (cf. *Gaudium et Spes*, 22). In other words, this is not only relevant to the possible salvation of all who will ever come into existence, it is most certainly relevant to all who are, on coming into existence, frozen or discarded as waste, and it is most certainly relevant to all who are aborted or lost through miscarriage.

Secondly, the word of God is ever alive and active and suitable for any good work (cf. Heb. 4; 12). As the salvation of human embryonic children is a matter of life and death, both for them and for those who put them to death, it is a 'good work' of salvation that impels us to draw upon the word of God. Therefore, as the aborted

---

[16] Cf. *Catena Aurea* by St. Thomas Aquinas, CHAP. 1 [of St. Luke's Gospel, From the comments on Lk 1:30–33], italics added: https://ecatholic2000.com/catena/untitled-62.shtml#_Toc384506962; and specifically, the GREEK EXPOSITOR. (Sev. Antiochenus.).

and the frozen embryo are embraced within the framework of redemption, then the word of the Lord can both inspire our understanding of what is involved and, indeed, help to bring about the love of our neglected neighbour. And, as we said at the outset, who is poorer, more powerless and of greater insignificance than the human embryonic child? And, therefore, who is more in need of the Good Samaritan's help than those threatened with abortion, experimentation, or being frozen indefinitely? And, therefore, how much more urgent is it that we draw upon whatever can be gleaned from the Scriptures to help with the recognition of the beginning of human personhood. Thus, if we can dissipate and dissolve the falsehood that these embryonic children are merely indifferent blobs of cells, let us do so and, in so doing, let us proclaim the good of life for all.

Thirdly, truth will not contradict truth, nor faith and science conflict where each identifies what is relevant to the benefit of the human embryonic child, whether newly conceived, aborted or frozen in the very throes of his or her natural development. Where reason and evidence proclaim the first instant of human fertilization to be the first instant of being a person, there is an obligation on all of us to ensure the good of completing human, embryological development. Thus it is necessary to end the profoundly unjust practice of the living taking the life of living, developing, embryonic children. Why not invest in the saving of life instead of the technical production of human embryos, entailing as it does multiple fertilizations, discarding or experimenting upon them or, finally, freezing "left over" human embryos? Why not invest in and investigate the possibility of the transfer of an ectopic embryonic child instead of the massive money-making project of destroying human fertility through unnatural and unnecessary contraceptives? Never mind the artificial arguments that spawned technological intrusions into the beginning of human life and, in the process, destroying many of the lives that would otherwise have wanted to thank you for your help?

At the same time there are those developing the possibility of "artificial wombs" that may, possibly, save the life of a child born prematurely at twenty-two to twenty-four weeks. Naturally there is a tremendous desire to see the premature child being helped to live. But, in these uncertain early days, there is a reality to the view that premature children will be doing a real experiment: an experiment that may or may not help them mature; that may or may not leave them crippled; that may or may not save their life[17].

But, conversely, will those who develop the technology go on and sell it to those who will use and abuse it, developing a technological womb for those artificially fertilized – not knowing what the psychological impact will be on those who are "manufactured" and not mothered? Already, it is reported, that "spare human embryos" are already "testing" what are, in effect, early attempts to manufacture wombs.

> What are human beings doing to "the" other: They 'first engineered endometrial cells – cells that make up the lining of the uterus – to create a number of microscopic uteruses.

[17] Lucy Tu, November 5th, 2025, "This machine could keep a baby alive outside the womb. How will the world decide to use it?": https://www.theguardian.com/world/2025/nov/05/baby-alive-outside-womb?CMP=oth_b-aplnews_d-5.

Human embryos from IVF centers were then placed in these uteruses, while the scientists watched whether implantation was a success or failure.

Of course, the embryos, whether they implanted or not, all soon died, sacrificed for "science"'[18].

And what about the complete, grievous irony, that while some members of the human race are at pains to preserve life, millions of human beings are being wasted, whether from conception or after? Is it all about money? Money from killing and money from saving lives?

If it once "seemed" as if research, using human, embryonic children, was unregulated; now it is clear that *it is* unregulated.

Where is the love?

---

[18] Steven Mosher and Chiara McKenna, January 19th, 2026, "Why the artificial womb is the next pro-life battleground: Human sacrifice, whether to ancient pagan gods, or modern medical science, is never acceptable": https://www.lifesitenews.com/opinion/why-the-artificial-womb-is-the-next-pro-life-battleground/

Chapter Four

A Closer Examination of the Biblical Text of the Annunciation

In this chapter there is a more exact and detailed reflection on "Who" is present at the Annunciation and how it can help us to read our relationship to the Blessed Trinity. This brings to light the presence of the Blessed Trinity at the "creation" of each embryonic human child, who then exists in-relation to the *Incarnation* of the Son of God. Furthermore, Mary's experience is also explored from the point of view of helping us to understand the possibility of a wife adopting a "product" of the *in-vitro* fertilization process, namely a hitherto frozen, now thawed, human embryonic child. For, just as none of us are outside a relationship to Christ, so there is an "overflowing" love, as it were, of His Blessed Mother, the Communion of Saints, the angels, and indeed the whole Church and the rest of mankind – all of which is called to focus on the least, most powerless among us, the embryonic child, whether he or she is waiting in death row or is about to be frozen.

Coming closer to the event of the Annunciation, we come into the presence of a most holy moment, both at the time intensely personal and now, no less so, but now made a gift to us all through the *Gospel of St. Luke.* Just, then, as there is a transformation of a uniquely personal moment into a biblical revelation of the *Incarnation* of the Son of God, so what has happened becomes a source of inspiration for our understanding of the love God expresses for each one of us. In other words, just as the moment of human conception is a profoundly personal event for the person who comes into existence, as indeed it is for his or her parents, so the more is known about this event the clearer it is that there is no beginning other than the beginning that each one of us is given: the beginning to exist. And, as we believe, there is no beginning to the eternal Son of God, yet there is a beginning in time to the humanity of Christ conceived under the heart of Mary.

We cannot, therefore, overlook the terribly poignant irony that those who, equally, have been given life, waste the gift of life that others have been given. As we proceed with our inquiry, it cannot fail to arrest our attention that just as Christ's conception was followed by an awful outbreak of violence (cf. Mt. 2: 16-18), so now that fear, given a different name, namely that of overpopulation, continues to destroy innumerable of the most powerless, poorest, and most vulnerable of our human family: embryonic and unborn children. In turning to the *Annunciation,* then, we are not looking for the data of the sciences, so much as the data of revelation. The *Annunciation* is a profoundly personal word which is imbued with an irreplaceably universal significance – not because of who we are but because the Holy Spirit has revealed the mystery of a woman becoming the Mother of God.

Let us then listen, following the course of Mary's dialogue with the angel Gabriel, thinking as we do: What has this to do with embryonic children, whether unborn or in a freezer? In this dialogue, Mary asks how it is possible that she can become pregnant if, while married to Joseph, she does not know a man:

> '[*"But how can this come about, since I have no knowledge of man?"*[1]] *And the angel said to her, "The Holy Spirit will come upon you, and the power of the Most High will overshadow you; therefore the child to be born will be called holy, the Son of God ...*"' (Lk. 1: 31-35)[2].

What follows, then, is in three parts: The verb to know; and the presence of the Blessed Trinity; and a Theological Commentary on certain aspect of the *Annunciation.*

Without, however, going too far into comparative texts or translations, it seems clear that there is a certain contextual contrast between *The Book of Genesis* and the *Gospel of St. Luke*: Adam was involved in the conception of Cain but Joseph was not involved in the conception of Christ. Adam was on the edge of Eve's dialogue with the snake but Joseph was not involved in Mary's dialogue with the angel Gabriel. But beyond these comparisons, both Eve and Mary, Mary more than Eve, can speak to our emerging theme of God's encompassing grace. Eve recognized the help of God at the conception of Cain; and, indeed, many others have cried out to God for the gift of a child. Mary, of course, is distinguished by her unique dialogue with the angel Gabriel and the wholehearted presence of God in the conception of her son Jesus.

---

[1] But I prefer *The New Jerusalem Bible* translation of 'no knowledge of man' of the Greek phrase, 'ou ginwoskw', as it is closer to the Greek and echoes Eve in Genesis, to be discussed later; *NJB*, London: Darton, Longman and Todd, 1985, italics added.

[2] *The Holy Bible*, Catholic RSV translation, 1966, italics added.

On the verb to know

St. John Paul II speaks of Mary and Joseph as married and remaining virgin: '*Although Mary is already "wedded" to Joseph, she will remain a virgin, because the child conceived in her at the Annunciation was conceived by the power of the Holy Spirit*' (*Redemptoris Custos*, 2[3]).' Indeed, *The Jerusalem Bible's* translation takes up the theme of virgin, even that of perpetual virginity, saying, "But how can this come about, since I am a virgin?"[4]

> The Catholic RSV translation, however, is a more interpretative translation, whereby in answer to the possibility of Mary becoming pregnant she says: "How can this be, since I have no husband?" (Lk. 1: 34)[5].
>
> Whereas *The New Jerusalem* text follows the Greek more closely, rendering the verb *to know* in a more ancient, biblical sense, of not having sexual relations with a man, "But how can this come about, since I have no knowledge of man?"[6]

---

[3] Italics added.

[4] *The Jerusalem Bible*, see also footnote, r.

[5] *The Holy Bible*, Catholic RSV translation, 1966.

[6] *The New Jerusalem Bible,* London: Darton, Longman and Todd, 1985; cf. also Etheredge, Chapter 1 of *Mary and Bioethics: An Exploration*, which looks at how the Christian sacraments arise out of the reality of the Holy Family, making the Holy Family a kind of watershed between Judaism and Christianity.

The latter translation, following the Greek words, ou ginwskw[7], more closely denotes the biblical sense, as I say, of no sexual relationship with a man or her husband. At the same time, the expression 'I have no knowledge of man' does not entail a denial of Mary's marriage to Joseph although it does express a specific denial of that intimacy more generally characteristic of marriage. Indeed, more than that, the Greek text and *The New Jerusalem* text "echo", by contrast, the words of Eve, Mary's precursor, as it were, of whom the narrator says, 'Now Adam *knew* Eve his wife, and she conceived and bore Cain' (Gn. 4: 1)[8]. The Hebrew word is, יָדַע, ya-da, which Brown-Driver-Briggs takes to mean, in this context, to '*know a person* carnally, of sexual intercourse'[9]. In another context, 'to know' can be understood to mean 'to know by experience'[10]. Thus we can take the more embracing sense that Adam and Eve have experienced an intimate knowledge of each other rather than the more reductive sense of 'sexual relations'. Indeed, as *The Navarre Bible* commentary on the *Pentateuch* says, 'the Bible uses the term "to know", thereby signalling the human depth of that relationship: although it takes place via the body, it does so in a context of mind and will'[11]. Furthermore, for those married couples who, like Tobias and Sarah, pray each

---

[7] *The Interlinear NRSV-NIV: Parallel New Testament in Greek and English*, Zondervan, 1993.

[8] *The Holy Bible*, Catholic RSV translation, 1966, italics added.

[9] *Bible Hub*: https://biblehub.com/hebrew/3045.htm: 3.

[10] *Bible Hub*: https://biblehub.com/hebrew/3045.htm: e.

[11] Editorial Committee, Jose Maria Casciaro *et al*, for the English edition, James Gavigan *et al*, Dublin: Four Courts Press, 1999, p. 58: Gn. 4: 1.

child into existence (cf. Tb. 8: 4-9)[12], they will know what an interpenetration of heart and spirit there is in the dialogue of marriage (cf. *Humanae Vitae*, 8-13, 15-16): a true interpenetration of both spousal union and being open to life[13].

At the same time, Eve goes on to say, emphatically, and without direct reference to Adam, "I have gotten a man *with the help of the Lord*" (Gn. 4: 1)[14]. Thus St. Luke, following Mary, takes up the theme of the action of God in bringing about the existence of the human being. In Mary's account, it is almost as if she is saying that while I do not know man, through the experience of the *Annunciation*, I know God's action in my life. On the one hand, then, there is Mary's intimate knowledge of God concerning the *Incarnation* of the Son of God. But, on the other hand, with respect to our theme of the relationship between the *Annunciation* and embryonic human life, there is both the ancient perception of the action of God at the beginning of human life and a radically new understanding of it in the light of the conception of Christ. We receive, then, from this discussion a kind *of priority of the action of God in the conception of human personhood; but not just an action of God – but an action of the Blessed Trinity.*

---

[12] Just as there is a dialogue between husband and wife, so there is a dialogue between them and God. Although, then, the text does not say, explicitly, that Tobias and Sarah came together as husband and wife, yet it is sufficiently suggestive that, given the holiness of spousal union, that it is fitting to take what they did as a prayer before coming together.

[13] Clearly 'being open to life' is a dialogue involving the whole spousal relationship and their circumstances and can be found more completely investigated in Etheredge, *Human Being: Moral Norm.*

[14] *The Holy Bible*, Catholic RSV translation, 1966.

On the presence of the Blessed Trinity

> 'The angel announced to … [Mary] not just the incarnation but fundamentally the entire mystery of the Trinity: "The Lord is with you" – that is Yahweh, the Father-God, whom she knows. Then to her wondering what sort of greeting this might be: "You will conceive a son" who at the same time will be the son of David. To her question how she should behave, since this son could not come from a man: "The Holy Spirit". The Trinity is therefore included in what befell her'[15].
>
> The following account emphasises Mary's response to the initiative of each person of the Blessed Trinity: 'Mary's cooperation in the work of salvation has a Trinitarian structure, since it is the fruit of the Father's initiative, who "looked upon the *lowliness* of his servant" (*Lk* 1:48); it springs from the *kenōsis* of the Son, who humbled himself by taking the form of a servant (cf. *Phil* 2:7-8); and it is the effect of the grace of the Holy Spirit (cf. *Lk* 1:28, 30), who prepared the heart of the young woman of Nazareth to respond at the Annunciation and throughout her life of communion with her Son'[16].

---

[15] Hans Urs von Balthasar's *Mary for Today,* San Francisco: Ignatius Press; 1st English edition (March 1, 1988), p. 35.

[16] From paragraph 6 of The Dicastery for the Doctrine of the Faith's document, 4th of November, 2025: "*Mater Populi Fidelis:* Doctrinal Note on Some Marian Titles Regarding Mary's Cooperation in the Work of

> Furthermore, we can take Elizabeth Rex's expression as, in a way, making these observations so much more explicitly trinitarian; she says: 'God the Father who created and infused … [Mary's] soul; God the Son who redeemed her; and God the Holy Spirit who sanctified her at the first instant of her Conception'[17].

We can see that the Blessed Trinity, the Father, the Son and the Holy Spirit, are present and active in the mystery of Christ's conception; and, if this is the work of re-creation, as it were, we can see it echoing the very first trinitarian act of creation. In the *Book of Genesis* there are three namings of God, each with the "surname" *Elohim*. The first time God is named, His name is *Elohim*, it is the bare name; but, however, this is a plural, dyadic noun, implying the presence of two: Is it credibly the masculine two of God the Father and God the Son? The Hebrew name for God, *Elohim*, however, is treated as a singular subject of the verb to create, which follows it, *bara*. Thus indicating that we are in the presence of a plural God acting as one in the action of creating 'the heavens and the earth' (Gn. 1: 1). The second time God is named there is a combined name which is feminine, *Ruach*, or Spirit, linked to the plural name of God, *Elohim* (Gn. 1: 2). Thus we have the expression, *Ruach Elohim*. The third time that God is named, this time in chapter two, God is called

---

Salvation": https://press.vatican.va/content/salastampa/it/bollettino/pubblico/2025/11/04/0829/01462.html#inglese.

[17] Etheredge and Elizabeth Rex, *The Zygote of Christ and the Mystery of Man*: Rex, Essay IV, p. 111: https://enroutebooksandmedia.com/zygoteofchrist/.

The Lord God (where "Lord" is a translation of a name too holy to be uttered – but which is coupled with the plural name, *Elohim*[18] (Gn. 2-3[19]).

If mankind, in the beginning, is conceived in the active presence of the Blessed Trinity, and if the human nature of Christ is conceived in the active presence of the Blessed Trinity, and Christ is in all things the same as man except for being without a human father's male seed and sin, then it follows that each one of us is conceived in the active presence of the Blessed Trinity. Indeed, reflecting on the 'divine "We"', St. John Paul II says:

> On the one hand, 'Before creating man, the Creator withdraws as it were into himself, in order to seek the pattern and inspiration in the mystery of his Being, which is already here disclosed as the divine "We". From this mystery the human being comes forth by an act of creation: *"God created man in his own image,* in the image of God he created him; male and female he created them" (*Gen* 1:27)'[20].

---

[18] Cf. Etheredge, *Scripture: A Unique Word,* "Chapter Eight: Creation: The Archetypal Action of God": https://www.cambridgescholars.com/product/978-1-4438-6044-4; and cf. St. John Paul II, *Familiaris Consortio,* 11.

[19] The expression, the "Lord God" is used throughout the account of the creation of man and woman, which completes what was omitted in the first account of creation in chapter one; and the Lord God continues to be used when Adam and Eve and the Serpent are addressed after the Fall.

[20] St. John Paul II, *Gratissimam Sane*, 6.

> On the other hand, and in particular, 'The Second Adam stamped his image on the first Adam when he created him. That is why he took on himself the role and the name of the first Adam, in order that he might not lose what he had made in his own image'[21].

Continuing, then, with the view of the presence of the grace of God reaching out to the least among us is the very dramatic truth of the real presence of the Blessed Trinity from the beginning of creation and from the new beginning of the *Incarnation* of the Son of God and, therefore, from the beginning of each one of us. To use a poor analogy, it seems as if the very nature of God is like a "field of grace-love" surrounding the embryonic child and, as it were, "inducing the possibility" of a Baptismal spring in each person's life. Clearly, however, owing to the necessity of actual baptism for each one of us, "inducing the possibility" of baptism is more like the impulse of God that arises out of, perhaps, the very sacredness of human life as each one of us comes from the creative action of God (cf. *Humanae Vitae*, 13, and CCC, 366). In other words, while God begets us with, as it were, an impulse of grace-love, imparting to us a kind of dynamic orientation to the possibility of baptism, this "dynamic orientation" is not baptism but a kind of impulse towards it.

---

[21] *The Catechism of the Catholic Church*, CCC, 359, quoting from St. Peter Chrysologus, *Sermo* 117; PL 52, 520-521.

Person "in" God and in reference to human beings

The word 'person' was used to denote the mystery of God as 'three persons in one God'; and, therefore, person goes beyond its original Greek meaning of mask (*prosopon*) to take on a meaning of "relationship". Person is a term that is used to communicate the mystery of the Blessed Trinity, in that not only is each Person wholly God – but that each Person is one of three '*subsistent* relations'[22]. While this can be a very difficult concept to grasp, it invites us to ponder that the three persons in one God, are each wholly God and are, at the same time, relationships that are '*internal* relations' to God: God the Father is irreplaceably Father to the Son; God the Son is irreplaceably Son from the Father; and God the Holy Spirit is irreplaceably from the Father and the Son or from the Father through the Son (cf. CCC, 248). In other words, in God "person" expresses "relationship".

In human terms, then, if we are made in the image and likeness of God, then there is a 'certain resemblance between the union of the divine persons and the fraternity that men are to establish among themselves in truth and love' (CCC, 1878, quoting *Gaudium et Spes*, 24). However, I want to draw attention not just to the 'resemblance between the union of the divine persons and [human] fraternity' but to the nature of being a divine person and being a human person. 'Society is not for ... [man] an extraneous addition but a requirement of his nature' (CCC, 1879). In other words, "relationship" is

---

[22] Brother Andre Marie, January 25th, 2008, "The 'Relations' in the Blessed Trinity": https://catholicism.org/the-relations-in-the-blessed-trinity.html.

constitutive of each one of us; although, it seems, it is also true that each one of us stands alone, as an individualized human being. It is in the nature of human being, then, to be both "an individual" and "in-relationship". So when a human being is conceived, he or she is conceived through "relationship". On the one hand a child is conceived "through" the relationship of husband and wife; and, on the other hand, in the act constitutive of conception God brings a child to exist: a person-gift: a person-gift from the reciprocal gift of husband and wife and the action of God.

The point, therefore, of dwelling on the mystery of relationship, both at the heart of each person of the Blessed Trinity and in the identity of each human person as daughter or son, is to show that the interrelationship of human identity and human relationships is so close as to indicate that when the Church says that human life is sacred because of the action of God, God's grace is already, as it were, transmitted to the child (cf. *Humanae Vitae*, 13). Just as husband and wife come together in the spousal union, constituting a unique intimacy from which human life originates, so the act of God which brings human personhood to exist transmits what can be called 'prevenient grace': the grace that disposes a person to die to sin and to enter 'into the life of the Most Holy Trinity through configuration to the Paschal mystery of Christ (cf. CCC, 1239). Prevenient grace is 'the grace of God in a person's life which precedes and prepares for conversion'[23]; and, it seems possible, this is precisely what is passed to the embryonic child as a consequence of being brought by God to exist and to be 'sacred': to be "set-aside" for God. And, in that all

---

[23] "Prevenient grace": https://en.wikipedia.org/wiki/Prevenient_grace.

grace tends to union with God and with one another in Christ, prevenient grace sets a precedent, as it were, in the very being of the embryonic child: a precedent of being disposed to seek God.

In other words, drawing on the adage, activity manifests being, there is a clear relationship between the mystery and action of God as the Blessed Trinity; however, even if it is true that the existence of the Blessed Trinity was not too explicit in the history of salvation[24], nevertheless the disclosure of it transcends the time of discovery and, therefore, applies to the whole of creation's relationship to God. In particular, then, we can consider there to be a relationship between the coming into existence of the human person and the "goal" of union with the Blessed Trinity. Just, then, as a child is "orientated" towards his or her parents, so God "orientates" us to Christ and the whole of the Blessed Trinity as to a door into the being of God.

What about the child not conceived in spousal union?

> *Dignitas Personae* says: '*the acts that permit a new human being to come into existence,* in which a man and a woman give themselves to each other, *are a reflection of trinitarian love.* "God, who is love and life, has inscribed in man and woman the vocation to share in a special way in his mystery of personal communion and in his work as Creator and Father" (9)[25].

[24] Cf. CCC, 238-248 and 290-292 etc.

[25] *Dignitas Personae*, 9:

Given, then, that the *'acts that permit a new human being to come into existence'*, in the instance of the reciprocal self-giving of husband and wife, are clearly absent if a child is brought to exist in a glass dish, could it be argued that there is no *'reflection of trinitarian love'* in what brings this child to exist? On the other hand, while not disagreeing with the embodied sign language of a husband's and wife's reciprocal self-giving, what is of concern here is that there is a divine-human *act of the child's coming to exist* entailing a relationship to the Blessed Trinity. Just as the act of the *Incarnation* of the Son of God entails an intimately inseparable relationship to the mystery of the Blessed Trinity so does this suggest that, similarly, each of us is conceived in the light of a relationship to the Blessed Trinity.

Moreover, just as each person is not a person in isolation but a human being-in-relationship, it follows that "we" are in a relationship to the Blessed Trinity "through" each other. Thus our response to each other entails, as it were, a response to the presence of God who brings each of us to exist in-relationship to His redeeming love. Each human embryonic child, however conceived, is conceived *in-relationship* to the creative-redeeming love of the Blessed Trinity. This relationship between God and each other implies an imperative that we act for the good of each other: an imperative already expressed in the natural law *to do good and to avoid evil.*

---

https://www.vatican.va/roman_curia/congregations/cfaith/documents/rc_con_cfaith_doc_20081208_dignitas-personae_en.html#_ftnref16.

A theological commentary on certain elements of the Annunciation

On the one hand, the *Annunciation*, the "moment" of the conception of Jesus Christ, the moment of the *Incarnation* of the Son of God, presupposes Mary's own, *Immaculate Conception*, a human conception identical with our own except for sin[26]. Furthermore, Mary's Immaculate Conception is a complete and total unmerited gift in view of her vocation to be the Mother of God; and, as such, Mary's own mystery epitomises the very nature of the human person as the recipient of God's gifts[27]. Thus Mary is both conceived without original sin and remains without sin and, therefore, transmits the flesh of the redeemed to her son, Jesus Christ[28]. On the other hand, as we know from *Gaudium et Spes*, 'by his incarnation, he, the son of God, has in a certain way united himself with each man' (22)[29]. This 'certain way', while being unspecified, could entail the

---

[26] Cf. Etheredge, *Mary and Bioethics*, 2020, "Chapter Five: The First Instant of Mary's Ensoulment", pp. 152-173; clearly, however, I am not including the creation of Adam and Eve, who were also created without original sin.

[27] Cf. 'When the Gospel explains that, in response, the disciple (who represents all of us) took her in, it uses a verb (*lambanō*) that, in this Gospel, carries the sense of "receiving" from faith (cf. *Jn* 1:11-12; 5:43; and 13:20).' From paragraph 6 of The Dicastery for the Doctrine of the Faith's document, 4th of November, 2025: "*Mater Populi Fidelis:* Doctrinal Note on Some Marian Titles Regarding Mary's Cooperation in the Work of Salvation": https://press.vatican.va/content/salastampa/it/bollettino/pubblico/2025/11/04/0829/01462.html#inglese.

[28] Cf. Etheredge, *Mary and Bioethics*, 2020, Chapter Five.

[29] Indeed, this text of *Gaudium et Spes*, 22, is drawn upon in the opening paragraphs of *THE HOPE OF SALVATION FOR INFANTS WHO DIE*

very moment that the *Incarnation* entered the Son of God into contact with human nature and, therefore, into contact with every person who exists.

It is possible, then, to understand a threefold mystery: that the Son of God has entered into a relationship with the whole human race in such a way that this relationship is the divine-human prerequisite of the grace of redemption for all. Any salvific divine-human act of salvation "builds upon" the foundational relationship that Christ entered into on being *incarnate* of the Virgin Mary. 'Therefore 'Let us dig deep, unhindered by worldly criteria, which too often demand immediate results and disregard the wisdom of waiting'[30]. Secondly, Mary herself is intimately implied in this relationship, both from the very first instant of her own conception and, if this is possible, through the explicit expression of this in the *Incarnation* of the Son of God that, concretely, entails his graced union with her. The graced union between Jesus Christ and His Mother Mary rests on the inseparability of both being free from original sin and both

---

WITHOUT BEING BAPTISED, paragraph 88; and, in numerous places, this document quotes the view that God can save, can communicate His saving, paschal mystery, in a way known only to Himself (also *Gaudium et Spes*, 22: 6, 31, 41, 81, 85, 93, 96 without, however, going into any nuances in the use of this reference). At the same time, however, this does not deny the necessity of the ordinary means of salvation; however, it does express the view that God is not restricted to saving us through His otherwise necessary sacraments of salvation.

[30] Zenit Staff, November 9th, 2025, 'Pope Leo XIV, "What does it mean to be the Church? Pope Leo XIV's three reflections (among which liturgy stands out)"': https://zenit.org/2025/11/09/what-does-it-mean-to-be-the-church-pope-leo-xivs-three-reflections-among-which-liturgy-stands-out/?eti=28540.

being intimately ordered to one another as mother and son. Thirdly, that each human being that comes into existence, comes into existence 'in a certain way united' to Jesus Christ. However, in the case of Mary, the spiritual union which both arises out of her relationship to her son, transcends the limitations of a natural, mother-son relationship. In other words Mary, in both being the Mother of the Lord, and redeemed by her son, enters upon a fullness of relationship between herself and her son which can only be described as both fully redeemed and ordered to the redemption of mankind.

In virtue, then, of who Christ is, and her *Immaculate Conception*, which was made possible by her son, Jesus Christ, the natural mother-son relationship is totally transformed, even transfigured, so that it expresses the wondrous universality of Mary being, in fact, the mother of all the living: the new Eve. But so hidden was this mystery in the relationship between Mary's *Immaculate Conception* and the *Incarnation* of the Son of God, that it was not revealed until Christ accomplished His definitive self-offering to the Father, which took up all that that entailed in His relationship to His mother. Thus the mother-son relationship was so opened through the mystery of the life, death and resurrection of Jesus Christ, that their relationship was universalized:

> 'When Jesus saw his mother, and the disciple whom he loved standing near, he said to his mother, "Woman, behold your son!" Then he said to the disciple, "Behold, your mother!". And from that hour the disciple took her to his own home' (Jn. 19: 26-27).

What we discover, in these translucent words, is that just as Christ 'in a certain way united himself to each man' so, simultaneously, He united 'each man' to His relationship to His Mother. Thus not only is He the redeemer of all mankind but she, in being His Mother, became the Mother of all the living. In other words, everyone who comes into existence, not only comes into existence in a relationship to Jesus Christ – but everyone who comes into existence comes to exist in a salvific relationship to Mary, His Mother. But, lest this is thought to make Mary our saviour, let us remember that the vocation of Mary is to say to us: "Do whatever he tells you" (Jn. 2: 5).

To recapitulate an earlier point, that owing to the act of the Creator at the beginning of each one of us, the life of each one of us is sacred; and, in being sacred, the life of each one of us is disposed, by prevenient grace, to choose the relationship into which our very existence is brought to exist. Just, then, as a child is born into the family which his or her parents have founded, so each one of us who comes into existence, comes into existence into the mother-son relationship, which implies all other human relationships, and so each one of us comes into existence in the family of God – disposed by prevenient grace to accept entering fully into this redeeming relationship of all relationships.

This is not proposing that anyone is with grace, on coming into existence, like Christ and His Mother Mary are, or even that Adam and Eve were; rather, it is proposing that we all come into existence disposed to be *in-relationship* to Christ and His Mother Mary. As St. John's Gospel says, we have been given grace upon grace: 'And from his fulness have we all received, grace upon grace' (Jn. 1: 16). The grace of being united to Christ, 'in a certain way', the grace of being united to Mary, His Mother, through being united to Christ, and the prevenient grace which disposes us to enter more fully into the very relationships into which our very existence has already entered us. And, just as we are born into a family of relationships, beginning with our immediate parents and grandparents, so we are, as it were, born "in view" of a relationship to the Holy Family and to the family of the Church and indeed to all mankind. Thus, just as every truth leads to *The Truth* (cf. Jn. 14: 6) so all relationships lead to Christ.

This said, it is clear that the ontological relationships into which we enter, on coming into existence, are of their very nature a gift of God – both totally gratuitous and foundational for our salvation. It is this relationship to the Son of God and His Mother Mary, established in the very act of the *Incarnation*, which builds upon Mary's *Immaculate Conception*, which is the very vehicle of grace, flowing as it does from the Son of God to us. Whether the grace-of-relationship is communicated through these initial gifts, the sacrament of Baptism or, in whatever way God pleases, in that He is not bound by His own sacraments, it is all from Christ and has as its goal to return us to Christ. We are just like the vine in the Gospel to which we are attached and, through which, God acts with us, or we act with God (cf. Jn. 15: 1-11); and, at the same time, this is the foundation for Christ's identification with us: 'Truly, I say to you, as you did it to one of the least of these my brethren, you did it to me' (Mt. 25: 40): the grace of Christ's union with us.

Chapter Five

A Discussion on Embryo Adoption in the Light of the Annunciation

We have touched on the plight of those children brought to exist in a glass dish. Therefore, it is necessary to consider them again in view of the possibility of their rescue, either from being "spare", "frozen", experimented upon or destroyed. One of the key problems is how to "see" that the human embryo, that is now being treated as a commercial property to be bought, sold, stored or destroyed – is not a commercial property but a child. What is more, these children are already in the context of an immense destruction of their contemporaries, either by surgical abortions or by abortion pills. So what will open us up to a fresh perception of these children's real reality? For they, along with the rest of mankind, are the only creatures God wanted for their own sake:

> This likeness [between the union of the divine Person, and the unity of God's sons in truth and charity] reveals that man, who is the only creature on earth which God willed for itself, cannot fully find himself except through a sincere gift of himself.

If God, in other words, so loves man, 'who is the only creature on earth which God willed for itself,' then who are we to enslave one another to destruction, to destitution, or to dehumanizing conditions?

Three considerations: The perspective of salvation-adoption; the Incarnation, Mary, and "embryo adoption"; and the equality before God of all and the gift of the child

The perspective of salvation-adoption

> 'From the lips of Jesus to the pen of Paul, the radical notion that God adopts us as true sons and daughters through baptism has been a core belief' ['see Matthew 6: 9-13 and Romans 8: 15'][1].

Adoption through baptism, first of all, follows on the ontological relationship established through the Son of God's *Incarnation* with each one of us, arising as it does out of the mystery that He 'has in a certain way united himself with each man' (*Gaudium et Spes*, 22). In other words, the sacrament of *Baptism* "builds" on the foundation of the fact that Christ, through His *Incarnation*, unites Himself in a certain way with each one of us. On the one hand, as members of the human race, we are one body. But, on the other hand, through Christ uniting Himself in a 'certain way' with each one of us we are immersed in His family. Spiritual adoption through baptism, then, follows on an ontological relationship – that each person is "caused" to exist by God and that being "caused" to exist is to be caused to exist more fully in-relationship to the *Incarnation* of the Son of God, His holy Mother Mary, and the whole extended family of the

[1] Whitney Anderson's, Fall 2019, University of St. Thomas, Master's Thesis, "Justice for the Orphan: the Ethics of "Embryo Adoption"", p. 44, biblical references are from the original text.

redeemed and, indeed, of all mankind. The human embryo, then, in whatever way he or she comes to exist, comes to exist *in-relationship* to Jesus Christ; and, if *in-relationship* to Jesus Christ, then to Mary and to the rest of us. *Baptism* is both a development and a fruit of Christ's redeeming love. However, in the perspective of the redeeming love of Christ, the intention of God is already made manifest in that He has also "caused" us to be sacred through His direct creation of each one of us imparting, as I have said, a grace disposing us to choose to enter more deeply into the relationships in which our existence immerses us.

This love of God which descends, as it were, to the beginning of each one of us, is a superabundant love: a love ever seeking to reclaim what was lost, deficient and unrecoverable by our own efforts or intention. In the perspective of the redeeming action of God, then, the human embryo, whether aborted, discarded as waste, experimented upon or frozen *is the exemplary beneficiary of a divine love of the lost – being the most unreachable and lost in the human terms of being either trashed or arrested in development.* What is striking, here, however, is that the act of the *Incarnation* is not a potentiality; rather, the act of the *Incarnation* brings to exist a potentiality that is there because of it. In other words, because of the *Incarnation,* because of the reality of the relationship between Christ and each person that comes into existence, *there is a relationship to be developed.* Therefore, the relationship-to-be-developed is already actualized by Christ's *Incarnation* and seeks to be unfolded and developed further.

The human embryo, whether aborted or frozen, having come to exist from the first instant of human fertilization, is in-relationship to Christ. Christ is not in-relationship to a stage of human development, as if we are talking about a pre-human entity; rather He is in-relationship to you and to me from the very beginning of being begotten, one in soul and body (*Gaudium et Spes*, 14). The whole Christ, from the moment of His *Incarnation*, is in-relationship to the whole human person: each is a human being-in-relationship. Moreover, relationships are reciprocal, although different between different members of a family – but all are called out of the closed circle of our own self-preoccupation to go out to the "other", whether that other be family, friends, spouse, neighbour or Christ and His Church. In contrast to the Pharisee who prays 'to himself' (Lk. 18: 11), we are called to enter into a relationship with the "Other" who is God and the "other" who is our neighbour.

The act of embryo adoption, like pregnancy generally, is not without its risks (Tobit 4: 4); but it is an act of gratuitous love *in the perspective of the divine-human adoption of man by God.* So, 'just as redemption goes beyond original sin without endorsing it, so an adopting love goes beyond the injustice of a child conceived "maternally homeless" without endorsing the method through which the injustice was perpetrated'[2].

The Incarnation, Mary, and "embryo adoption"

Is it significant that Mary is alone when she accepts the possibility of the *Incarnation* of the Son of God? Is it significant that Joseph

---

[2] Etheredge and Elizabeth Rex, *The Zygote of Christ and the Mystery of Man*: Etheredge, Essay III, p. 65. https://enroutebooksandmedia.com/zygoteofchrist/.

had no say in Mary's acceptance of becoming the Mother of the Lord? Mary is betrothed to Joseph but decides, independently of him[3], or so it seems, to cooperate with God and to accept conceiving the Son of God 'made flesh' (Jn. 1: 14); however, what seems to be a decision independent of Joseph, only seems to be so, superficially.

It is the will of God that Mary and Joseph are betrothed. It is the will of God that they do not come together as man and wife. It is the will of God that Mary remains a virgin, while virginally conceiving the Son of God in her flesh. It follows, then, that it is not a betrayal of Mary's marriage to Joseph that she has accepted this singular but complex vocation. Indeed, this is confirmed by God revealing to Joseph that the child she has conceived is uniquely of the Holy Spirit and that, therefore, he can accept Mary, pregnant as she is, into their home (cf. Mt. 1: 18 – 25). In other words, Mary's action, in so far as it is in the *perspective of the will of God* is, in reality, neither contrary to the nature of marriage nor an offence to Joseph. What enables Joseph's acceptance of this mysterious conception of Jesus is the fact that he, too, has undertaken his vocation to be betrothed to Mary, in the same *perspective* of doing the will of God.

What, then, seems to be the question, here, with respect to the adoption of the frozen embryo if not the following: Is it within the will of God? Is it the will of God that this child, conceived in a glass dish, frozen, thawed, is then transferred to the womb of a woman

---

[3] Whitney Anderson's, Fall 2019, University of St. Thomas, Master's Thesis, "Justice for the Orphan: the Ethics of "Embryo Adoption"", p. 35: https://ir.stthomas.edu/cgi/viewcontent.cgi?article=1022&context=sod_mat.

who is willing to mother him or her? Is it the will of God that this embryo adoption is a good moral act?

Ordinarily, one might say, there is the husband's and wife's complementary intention to be open to life. In the case of Joseph, his openness to life is evident in his acceptance of the conception of Christ within his marriage[4]; however, even more ordinarily, in the case of husband and wife being open to life, there is no certainty that a child will be conceived. Therefore, in reality, both husband and wife are, ordinarily, in the same *reality-relationship,* as it were, as Mary and Joseph. Given that there is no right to a child, Mary and Joseph stand in the same relationship to Jesus as all parents do to their children, namely, that all children are a gift to their parents: an unmerited gift to their parents. But, within the mystery of the will of God, everyone is loved into existence by God and entrusted to themselves and to their parents as a love-gift. Similarly, in the case of a child adopted as an embryo, if there is at least agreement that it is the will of God to accept this child into their marriage, then adoption is an expression of the love of God for that child. Moreover, without this adoption, this child will eventually atrophy and die, even if it takes longer than the longest child conceived from frozen, namely, having been frozen for thirty-one years![5]

---

[4] Cf. Francis Etheredge, "A Reflection on the Language of the Body": https://www.communio-icr.com/articles/view/a-reflection-on-the-language-of-the-body.

[5] By Julio Tudela and Cristina Castillo of the Bioethics Observatory, Published On: September 16th, "Baby born after 31 years of embryonic cryopreservation".

So what determines it to be the will of God that a child is adopted from being fertilized outside his or her mother's womb? The principal determinant is that God Himself has united Himself to the embryonic child, even as he or she is fertilized in a glass dish; and, therefore, what God has welcomed, let not man reject. And, as we have seen throughout this book, everything God does is from the point of view of rescuing each one of us, but not in isolation from Himself or from each other. Given, then, that there cannot be a more abandoned child, after that executed by abortion, than the one being frozen, it follows that the very relationship to God, Mary, and their extended family, is the prevailing motivation for taking up the child abandoned to freezing – even if there is no guarantee that the child will survive being thawed and transferred to a mothering woman's womb. For, left in the freezer, there comes a point when ultimate destruction follows anyway. Rescuing a frozen child is the complementary grace to Christ's act of redeeming love!

The morality of the transfer of an embryonic child to the womb of a willing married woman

Further to the question of a child being brought to exist in a glass dish is the question of the morality of transferring that child to a woman willing to accept him or her.

Although, then, this is a biblical answer to the question of the morality of embryo adoption, what is involved is not the text as if it stands separable from the reality it depicts; but, rather, the biblical text is an expression of the realities to which the words refer (cf. *Dei*

Verbum, 8)[6]. And it is the realities to which the biblical text refers that is the burden of this little book.

To begin with, then, there seems to be a starting point with the action of God: an action from ancient times understood to be present in the conception of a human being: an action, in view of the perfection of God, which has no other object than that of the good of the person who exists for his or her own sake[7]. Just, then, as God's embracing love of His creation includes the redemption and sanctification of human beings, so the spousal love of spouses unfolds in the different but complementary parental love of their children. Thus human beings love in their totality, like God, and not just in a functional sense as if offering the hospitality of the woman's womb can, in reality, be separated from her total identity as a person called to love[8]. Indeed, just as 'in marriage the heart of the beloved realizes that no other heart is capable of making it feel "at home" like that of the person it loves'[9], so how desperately true is the suffering brought about by a mother's tragic unwillingness to be the "home" of her

[6] Further discussion of the nature of Scripture is beyond the remit of this Chapter; but cf. Etheredge, *Scripture: A Unique Word*: https://www.cambridgescholars.com/product/978-1-4438-6044-4.

[7] Cf. Melissa Moschella, in "Protecting the Dignity of the Child: Understanding the Rationale for the Inseparability Norm in the Context of the Embryo Adoption Debate" in *Human Embryo Adoption, vol. 2, Catholic Arguments For and Against*: and cf. also, *Gaudium et Spes*, 24.

[8] Making more explicit the totality of the person implicated in parental love derives from reading the chapter by Christopher Tollefsen, "The Procreative Dimension of Marital Union and Embryo Transfer" in *Human Embryo Adoption, vol. 2, Catholic Arguments For and Against.*

[9] *Una Caro*, paragraph 140.

child. Conversely, let us recognize the generosity of a woman's self-gift when she takes the risks she does in doing the good she chooses to do[10]; and so let us listen to the words of Tobit: "Remember, my son, that ... [your mother] faced many dangers for you while you were yet unborn" (Tb. 4: 4).

Thus we can see that there is a natural emphasis on the relationship of the child to the mother – not only a maternal emphasis but almost a salvifically maternal relationship (cf. also Jn. 19: 25-29) in that the woman is particularly capable of recognizing that she is *in-relationship to God, the Author of life* (cf. also 2 Mac. 7: 1, 20-31, 39-42). Furthermore, whereas the *Incarnate Word* takes the flesh of the Virgin Mary, the adopted embryo, already formed as an embryonic human being, is given both hospitality and the whole human, mothering relationship, that enables the child to receive all that is entailed in being gestated beneath a mother's heart.

What has emerged, however, is not so much a discussion about the details of embryonic existence or transfer[11] but, rather, a kind of embracing circularity by which embryo adoption, in its holistic reality, contributes to rectifying the relationships from which the child has been unnaturally estranged. On the one hand, the child has been deprived of the naturally dynamic engagement with the embryological growth that is integral to his or her living response to life. And,

---

[10] Cf. Cara Buskmiller's excellent "Conclusion" from the chapter, ""Medical Aspects of Embryo Adoption" in *Human Embryo Adoption, vol. 2, Catholic Arguments For and Against.*

[11] Cf. A number of the position discussions in Anderson's, Fall 2019, *University of St. Thomas*, Master's Thesis, "Justice for the Orphan: the Ethics of "Embryo Adoption"".

on the other hand, the child has been estranged from the mother's recognition of her vocation to take up this child *into a relationship with the Author of life, to whom, as it were, she will introduce her child.* In general, the mother stands in a certain, more direct relationship to the action of God at conception, than her husband and, in view of this, she both introduces him to that action so that, together, they can bring the child up in *the perspective of the relationship to God that has been established.* Just, then, as God acted in the life of Eve, after the fall, drawing the good of life from Adam and Eve's knowledge of each other, so God draws the good of human life out of the real, however imperfect, beginning of a child being fertilized in a glass dish:

> 'St. Thomas explains that … "God allows evils to be done in order to draw forth some greater good. Thus St. Paul says, 'Where sin increased, grace abounded all the more' (Rom 5: 20); and the *Exultet* sings, 'O happy fault, … which gained for us so great a Redeemer'"'[12].

Therefore, whatever the circumstances of a child's origin, and however clearly it is contrary to our being 'the minister of the design established by the Creator' (*Humanae vitae*, 13), God strives to bring about the whole relationship to Him, and to others, to which each one of us is called in virtue of being members of the human race. In this instance, then, embryonic transfer, the transfer of an abandoned

---

[12] St. Thomas Aquinas, *Summa theologiae*, 3, 1, 3 and 3; cf. CCC, 412, quoted in *The Navarre Bible: Pentateuch*, p. 55.

child to the womb of a willing mother, is a totally different kind of act to that of fertilizing a child in a glass dish and transferring him or her to a "customer".

Rescuing a child that has been abandoned at the bottom of a cliff with the rope that was used to drop the child down there, is a totally different kind of action to that of the adult who abandoned the child at the bottom of the cliff.

The Marian Mission of the Church: "Mary arose and went with haste" (Lk. 1:39)[13]

This is, in part, the title of Pope Francis' address to the Youth of the World whom he met in Lisbon, Portugal, in 2023. What, then, is the connection between his address and this chapter on the *Annunciation* and embryo adoption?

> Mary 'got up and went forth, into the world of life and movement. Even though the astonishing message of the angel had caused a seismic shift in her plans, the young Mary did not remain paralyzed, for within her was Jesus, the power of resurrection and new life. Within herself, Mary already bore the Lamb that was slain and yet lives. She arises and sets out, for she is certain that God's plan is the best plan for her life. Mary becomes a temple of God, an image of the pilgrim

---

[13] Pope Francis, « The Marian Mission of the Church": https://press.vatican.va/content/salastampa/it/bollettino/pubblico/2022/09/12/0674/01382.html#inglese.

Church, a Church that goes forth for service, a Church that brings the good news to all!'[14]

Mary, then, was touched by the possibility of her elderly cousin's needs, to whom God granted a child in her later years; and, therefore, she 'arose and went with haste'. Thus the question for us is this: Do we arise and go with haste concerning the needs of the aborted and frozen human embryos? Are we doing everything we can to resolve this question of whether or not it is right to intervene on behalf of the forgotten, aborted, and frozen, some of whom have been frozen for thirty years and lived?! Are we advancing the charity necessary to save the world and advertising the organizations which help to make known the possibility of adopting a frozen embryo – even if the Church's uncertainty is, at the very least, a window of opportunity?

All in all, then, it is fitting to call this account of the "Saving of Embryonic Children" Marian, as it focuses on the existence of the child first and foremost and it is about seeing the person 'with … [our] hearts'. In a word, when a mother mourns the loss of a child through miscarriage, it is not a mourning that is meaningless. It is precisely because the mother sees with her heart the loss of a person: her son or daughter; and, at the same time, remembers that he or she has gone before her to *The Father.* In a word, then, there cannot be a full account of a mother and father's experience[15] without the

---

[14] Pope Francis, « The Marian Mission of the Church".

[15] Cf. Francis Etheredge's, *The Prayerful Kiss*, and the prose and poem called "Indelible": https://enroutebooksandmedia.com/theprayerfulkiss/;

recognition, in all the highways and byways of bioethics, that it is not "what" begins – but "who" begins at conception or fertilization. Conception, however imperfectly it occurs, occurs within the woman who becomes a mother and fertilization, while occurring in the former is stripped of its welcoming humanity when it occurs in a glass dish. Clearly, though, this is about the beginning of a child. And, indeed, without a "who" being either conceived or fertilized, then there is no union, in a certain way, to the *Incarnation* of Christ (cf. *Gaudium et Spes*, 22). Therefore, whether frozen or not, if Christ united Himself in a certain way with each human being at conception or fertilization, then we need to follow His lead too, to the depths of human existence and the completing human development which follows on embryo adoption and rescue.

Perhaps we need to let this plight of the unborn, of the aborted, the experimented upon and the frozen human embryonic person, to reach us: to let it be like a sword through our heart (cf. Lk. 2: 34-35). Perhaps we need to let the plight of the unborn help humanize us and focus us on "going out to the other" – reaching out to the other who is our brother or sister and who implicates us, as it were, in their continued and prolonged suffering[16]. Perhaps we need to hear the words of Christ addressed to us: "What you did to the least, the most powerless and unimportant, you did to me" (cf. Mt. 25: 40).

Our equality before God and the gift of the child

---

and, indeed, there is also Sean Corcoran, "Men for Life": https://menforlife.org/

[16] Cf. Joshua Hordern, "'A Knife into My Heart': Cries, Compassion and Ethical Life": https://www.tandfonline.com/doi/full/10.1080/20502877.2022.2124604.

In reality, then, the very dependence of parents on God's gift of a child *puts all parents on an equal relationship to both each other and to Mary and Joseph.* Therefore, in the *perspective of the child-as-gift*, the child fertilized in a glass dish, frozen, thawed and transferred to the womb of the woman-who-is-willing-to-be-mother is no less a gift than any other child: a gift to both those who were the "originating" couple, by whatever means, and a gift in the nature of being in existence through a gratuitous act of God. In other words, irrespective of how a child comes to exist, the existence of a child is always and everywhere a gratuitous act of God's generous giving of existence; and, as such, an expression of the person being an irrevocable end in themselves: 'man is the only creature on earth that God has wanted for its own sake' (*Gaudium et Spes*, 24). Moreover, 'if man is the only creature on earth that God has wanted for its own sake, man can fully discover his true self only in a sincere giving of himself [cf. Lk. 17: 33]' (*Gaudium et Spes*, 24).

If, then, the nature of a person as self-gift is inscribed in the very existence of the person-as-gift, then it follows that the nascent nature of the child "calls out", as it were, to complete the cycle of being given existence by making a gift of his or her existence to God. Parents, then, who are themselves 'ministers' (*Humanae vitae*, 13) and not masters of the gift of children, are both co-creators and co-operators with the work of God. Therefore, this natural characteristic of the very nature of the person, to make a gift of him or herself, is an implicit appeal to the human family – first to the natural parents but also, in and through them, to the whole human family. Whomsoever answers this appeal, whether it be the natural parents or others, are answering this call to be "in-relationship": to do the completing good which the human embryonic child's life literally depends upon.

Overall, then, while there are significant differences with respect to the *Incarnation* of the Son of God and embryo adoption, there is a common language of the *person-gift* of the child. At the same time, there is a common orientation of parents to the acceptance of the *child-as-gift*. And, if the parents cannot or do not fulfil their vocation to welcome the child, then it falls upon other members of the human family to do so. This is both because of the right of a child to be welcomed as a gift and to the completion of his or her development. This seems to suggest that adopting a human embryo falls within the *logic of salvation history.*

The very different reality of the *Incarnation* leads us to a reflection in *favour of embryo adoption as the expression of the will of God, in that God is already disposing the child to seek Him. Thus rescuing an abandoned human embryo is a kind of "incarnate" expression of the redemption which impels God to help each one of us on coming into existence. Rescuing an abandoned human embryonic child finds its fullest expression in fulfilling the law of charity: of fulfilling the love of God and our neighbour as ourselves. If God does not "drop the child cold", without hoping to start him or her on a lifelong search for Him, then we must pray to give him or her a warm welcome too.*

The Implicit Definition of Human Conception

On the one hand, it can be argued:

> 'A new human life begins at the moment when the genetic information contributed by the sperm from the father is combined with the genetic information contributed by the ovum (egg cell) from the mother. As soon as fertilization is completed, a new human being begins its life'[17].

The emphasis of this definition is on the completion of a stage of development, namely, fertilization; as it says: 'As soon as fertilization is completed, a new human being begins its life'. The aforementioned account, then, supposes a type of delayed animation and suggests, although it does not explicitly state, that ensoulment occurs after 'fertilization is completed' because, it then says, 'a new human being begins its life'. However, the life of the embryonic human child began at the first instant of fertilization, when "life" was actually transmitted.

In contrast to this idea of delayed animation, The Latin edition of *Donum Vitae*, The Gift of Life, says at footnote 27:

---

[17] "Keys to Bioethics": http://www.laityfamilylife.va/content/dam/laityfamilylife/amoris-laetitia/KeystoBioethics/Keys-to-Bioethics-EN-1.13_SITOWEB.pdf.

> 'Zygotum est cellula orta a fusione duorum gametum'[18]: the Zygote is the cell arising from a fusion of the two gametes[19].

In other words, if ensoulment entails the fusion of the nuclear content of the sperm and the egg-cell, then it follows that ensoulment cannot be the first instant of fertilization. And, if ensoulment is not the very first instant of fertilization, then at what moment did Mary's *Immaculate Conception* occur? For Mary's *Immaculate Conception* had to occur from the first instant of fertilization, if she was to be wholly human and wholly without sin. And, as St. Thomas Aquinas says, grace needs a subject; and, therefore, Mary had to be ensouled from the first instant of fertilization for her to be *Immaculately Conceived*, whole and entire, from the very beginning of her existence. Thus, as I have argued elsewhere, if the first instant of fertilization "originates" the embryonic life of the child, and the human soul is the "life" of the body[20], then it follows that the first instant of fertilization is the first instant of human personhood. This is because the first instant of the new human life is expressed by the closure of the embryonic wall and the fact that this is the first existential

---

[18] *Donum Vitae*, Latin text: http://www.vatican.va/roman_curia/congregations/cfaith/documents/rc_con_cfaith_doc_19870222_respect-for-human-life_lt.html.

[19] This translation of the Latin is deemed 'correct' by the Rev. Dr. Richard Conrad, OP, of the Oxford Dominicans (email, 17th of September, 2025); Fr. Richard is an Official Fellow, Lector, Member of the Aquinas Institute, Oxford.

[20] Cf. Etheredge, Chapter 12, "Life from Life", of *Scripture: A Unique Word*: https://www.cambridgescholars.com/product/978-1-4438-6044-4; and cf. CCC, 363.

existence of the human embryonic person (and all that that event entails). The very existence of the embryonic child's body is expressed by the enclosure of the embryonic wall and, therefore, this is the "outward" sign of the "inward" act of God which brings the whole person to exist from the very first instant of fertilization.

This conclusion is relevant to the whole discussion because it is this "moment" of human beginning which founds Mary's own, *Immaculate Conception*, the unity between each human person and the *Incarnation*, and thus to all whom Christ seeks to save. Just as at the conception of each human embryo, so at the fertilization of each human embryo, both are taken up into Christ's redemptive act as expressed by the *Incarnation uniting the Son of God, in a certain way, with each human being*. Therefore, the married woman who accepts an embryonic child, transferred to her womb, is acting in accordance with the graced good to be done to the thawed, nascent embryonic human life.

An End Word

Natural Truth and Law; Theological Truth and Salvation

Investigations which seem to be so different are in fact related in view of the integrity of creation which arises out of its single author, namely God, who created 'the heavens and the earth'. Thus our thoughts run in two directions, simultaneously, that of natural truth and law and that of theological truth and our salvation. What is more, given the gravity of the protection or destruction of human life, we are either gravitating towards salvation or away from it; but, either way, the more we ponder the discoveries of science and the mysteries of revelation, the clearer they assist and illuminate each other.

Natural truth and law

We live in a time when technology has, in part, proceeded independently of the moral law and, indeed, has therefore transgressed it. Thus, however imperfect, partial and patchy are the legal protections of embryonic children, we nevertheless need to keep examining what is now known for certain about the origin of human personhood and, as with any instrument of law, constantly challenging it to express and embody the rule of law: "do good and avoid harm". For we are not talking of an abstract entity, "technology", we are talking of real people making real inroads on the gift to humanity of man being male and female, called to marriage and to be open to life.

Moreover, where there is the problem of infertility or other obstruction to the human transmission of life, it is right and just that there is a sensitively human investigation of its causes and not a technological circumvention of what they are. If, indeed, there is a good remedy that is available to the husband and wife who are afflicted by this suffering, then let it be discovered and its balm applied; but, otherwise, let there be a consideration of another service to life if the obstacle to children being conceived within a particular marriage cannot be resolved in accordance with the dignity of both parents and child. For otherwise, as we shall see, a contradiction undermines the very good to be done.

The Trump administration, having expressed the truth of human personhood from conception have now contradicted themselves by incentivizing recourse to in-vitro fertilization.

> When did Trump's administration demonstrate the truth of human personhood from conception? When his Executive Order said that: '(d) "Female" means a person belonging, at conception, to the sex that produces the large reproductive cell. [And that] (e) "Male" means a person belonging, at conception, to the sex that produces the small reproductive cell'[1].

---

[1] President Donald J. Trump, January 20th, 2025, "Extreming and Restoring Biological Truth to the Federal Government": https://whitehouse.gov/presidential-actions/2025/01/defending-women-from-gender-ideology-extremism-and-restoring-biological-truth-to-the-federal-government/.

At the same time the Trump administration brings this reality of there being two sexes, that of male and female, to the United Nations, objecting to the use of undefined and unscientific terms which are also irrelevant to a document on the protection and promotion of the rights of women and girls[2]. Furthermore, 'The Trump administration issued new massive restrictions on U.S. foreign assistance that will end all direct and indirect funding of abortion, diversity, equity, and inclusion (DEI), and gender ideology overseas'[3]. These are visible steps towards advancing an international pro-life culture.

When did Trump's administration contradict personhood from conception? When it was said that 'President Donald Trump announced that his administration has struck a deal with pharmaceutical companies that will drastically reduce the cost of in vitro fertilization (IVF), triggering an outcry from pro-life and religious leaders who envision a vast increase in the number of embryos – unborn children – who will be treated as nothing more than medical waste as a result'[4]. In other words, the nature of the child, of the human

---

[2] Stefano Gennarini, J.D. and Iulia-Elena Cazan, "US to UN Women: Stop Pushing Gender Ideology": https://c-fam.org/friday_fax/us-to-un-women-stop-pushing-gender-ideology/.

[3] Lisa Correnti, "Trump Ends All Foreign Abortion/Gender Spending": https://c-fam.org/friday_fax/trump-ends-all-foreign-abortion-gender-spending/.

[4] Doug Mainwaring, October 17th, 2025, "Trump announces deal to make IVF easier to obtain, sparking pro-life outcry": https://lifesitenews.com/news/trump-announces-deal-to-make-ivf-easier-to-obtain-sparking-pro-life-outcry/

person, as a person-gift, is being supplanted by an ideological claim that there is a right to a child; but if a child, if a human person is inherently a person-gift, no ideological claim to a right to a child is sustainable.

> "IVF from beginning to end is filled with child sacrifice. From murdering innocent babies, to freezing them on ice, to eugenically choosing which ones to save, to separating them from their mothers to give to lgbt couples," said Lizzie Marbach, former director of communications for Ohio Right to Life. "It's a house of horrors and should be criminalized, not incentivized"[5].

Why does it matter that statements contradict each other? Because true statements go beyond themselves and express an unchanging reality. In other words, truth is not "atomistic"; truth is not unrelated to truth. In fact, just as a real atom forms a solid structure like a tetrahedron, so truth expresses the real structure of reality. If it is true that a human person begins at conception, then it cannot also be true that we can delete, discard, experiment upon, or freeze anyone who begins at conception or fertilization. If it is true that a human being is a person-gift, then it cannot also be true that the person has no rights; for if we are all person-gifts, and it cannot be otherwise, then we are all equal before God and the law. Furthermore, if a government recognizes the truth concerning the moment

---

[5] Doug Mainwaring, October 17th, 2025, "Trump announces deal to make IVF easier to obtain, sparking pro-life outcry".

a person comes to exist, then it does itself teach its citizens, and other countries, exactly that truth. Therefore, just as we do not invent the truth, let the truth determine how we live.

More recently, however, President Trump has said, in his message of January 22nd, 2026, for the "National Sanctity of Human Life Day "On July 4, 1776, our Declaration of Independence righteously affirmed that every human being is endowed by Almighty God with the unalienable right to life'; and again: 'Today, we uphold the eternal truth that every human being is created in the holy image and likeness of God, blessed with infinite worth and boundless potential'[6]

May the Lord grant consistency to this wonderful endorsement of the right to life by the Trump Administration[7].

Theological truth and salvation

In view of the inseparable interrelationship of the conception of Mary, not forgetting her Son, Jesus Christ, the Son of God, it has been necessary to revisit previous investigations so that we can see clearly how the Dogmas of the *Immaculate Conception* of Mary and the *Incarnation* of the Son of God can inspire, anew, the investigation of the beginning of each one of us. Because, after all, with the

---

[6] "Presidential Message on National Sanctity of Human Life Day, 2026": https://www.whitehouse.gov/briefings-statements/2026/01/presidential-message-on-national-sanctity-of-human-life-day-2026/.

[7] Cf., too, 24th January, 2026, "The Trump admin. must stop undermining pro-life culture": https://www.bibliatodo.com/En/christian-news/the-trump-admin-must-stop-undermining-pro-life-culture/.

deafening unwillingness of people to recognize the reality of nascent human life, we need to appeal to God Almighty to have mercy upon this generation, not only because of our sins, but also because of the ever-pressing plight of our unborn, embryonic children. Thus, because it is clearer than ever, this book has hoped to demonstrate that just as we need to rediscover the good of human life, so we need to rediscover that our salvation may depend, more than we realise, on recognizing the plight of the unborn as integral to our own salvation: that we cannot imagine, if ever we tried to, that our salvation is independent of the salvation of the "other". In other words, that if we do not love the least among us, then we cannot claim to love God; and if we do not love God, then we risk losing, too, our own salvation[8]. And who is the least of the least among us if not the aborted, abused and frozen embryonic child?

Therefore, we could well refer to this usurpation of technology's true service of man as an idolatry of technology that expresses the pride of man – as if there is no god greater than what man can devise (cf. Gn. 11: 1-9). And because "technology" is not a person but a tool of people, it is clear that there are those who both invent and use this technology as if it is not engaging in the destruction or manipulation of real children. But again we have a contradiction: If real children were not involved then the whole abortion and *in vitro* industries would collapse.

---

[8] Cf. Pope Leo XIV, *Dilexi Te*: "On the Love for the Poor": https://www.vatican.va/content/leo-xiv/fr/apost_exhortations/documents/20251004-dilexi-te.html.

But, conversely, God, as we know from the Scriptures, can humble the great of the earth, whether it be Nebuchadnezzar or a modern man. By contrast, then, the Church is rediscovering the mercy of God, pre-eminently expressed in the life of Christ and, increasingly, permeating the life of His Church. Simultaneously, there is a renewed understanding of the mystery of man, male and female, in the light of the *Blessed Trinity*: the three Persons in One God; and, therefore, a renewed investigation of the depth of human personhood, both in its own right and in the mystery of marriage.

Let, then, this truly vexed question of how to love the embryonic human child be undertaken afresh. Let us learn again how to love each child, his or her mother and father and their families too. Let us learn again how to love each one, whether conceived "to be aborted", fertilized in a glass dish, threatened with extinction, or the frozen unborn that need to be freed from the uncertainty as to whether their rescue can be welcomed by Holy Mother Church or not. Let the Church, then, approve the possibility of adopting thawed human embryos. Bearing in mind that this approval is a recognition of the needs of the frozen, human embryo, and not even an incidental endorsement of the injustice which brought about this frustration of the normal, human development of an actual boy or a girl. And let this recognition of the rights of human embryonic children re-grow, re-found, re-recognize the right of all unborn and incompletely developed children to precisely that: to the natural, human right of completing human development.

Finally, in the light of Mary, the Mother of the Lord's maternal care for all of us, let her intercede for all the unborn in view of the many and various ways that the most vulnerable among us are subject to every kind of technological abuse, from abortion to being frozen. But, just as Mary's Motherhood of God and of the Church has been recognized over the centuries, let her motherhood of us be more and more a reason for listening to the words of her Son: 'Let the little children come to me' (cf. Mt. 19: 14). Let, then, the prayer of the Church be raised, more and more, for the vulnerable, embryonic child, whether threatened with extinction or frozen!

Freezing or Thawing

Entering into time, Lord, you came among us to heal, to cure, to Transform, to inspire in us the belief that you are the Son of God and that the power of your word is a power which enters the heart to answer our deepest questions: questions which address who we are, who you are, and what you came to reveal about God's love.

So we discover, through Mary's dialogue with the angel Gabriel, that you, Lord, are the Son of the Most High and that the Holy Spirit has the power to 'overshadow' her and bring you to exist in the flesh; and, because of this dialogue we, too, are invited to draw Near to the Blessed Trinity and to bring all our heart's poor loving.

And thus, Lord, we bring all those who we suffer with, whether we know them personally or not, but especially if they are in our family, among our neighbours, on the street, in a doorway, sleeping under a tree, near a library, in a tent on the side of the street, or lying, almost unrecognizably, in the rubbish, hidden in the dirt.

But there are others, Lord, whom you know are out of sight but not beyond your penetrating gaze, seeing right through the closed Freezers, at temperatures where, ordinarily, no one lives, yet live the smallest, frozen unborn, barely in existence, but exist they do, as "To be" is irrevocable, and who will cry out for their right to grow?

While, then, these that are frozen are fortunate in comparison to those who have been "deleted", "discarded", "experimented upon", they are many and, frequently, even if they are thawed, they may never Survive and find a maternal home or, if they do, they may fail in the reviving of life, in "restarting", and die a thawed, withering death.

How many of us, Lord, while not exactly the frozen human embryos, buried alive in freezing nitrogen, are yet frozen out of society in so many different ways, as if there are those who go about with ice-cold Capabilities, shooting darts of addiction to drugs, alcohol, and pictures, preying on the prayerless, wordless, dying bodies of people?

When you came, Lord, it was your word, your touch, your company, which touched so many and, even if we deserted you, yet you came to save, to stay among us, and to forgive; and, having risen and ascended from us, yet you have stayed, silently, in our Churches, Being there, beckoning us to come to the warmth of your presence.

It is true, Lord, that we live in a world of contradictions where there are immense technological aids for those whose thoughts are within, being able to speak through a machine, or write, or in some way convey what they are thinking through the immeasurable help of what changes actions, or eye movements, into words out of silence.

But it is also true, Lord, that there is an immense wastage of human Life, like a slope of human slippage, sliding into the drain an Unknown, almost inestimable amount of human remains, but not just the remains of adults – but the remains of culled, cut, and incredibly Innocent human life, known in all his or her splendour only by you.

Only you, Lord, can call upon the heart of each one of us to come to The invisible fire of your Eucharistic presence to be soaked in the Mystery of a love that always seeks to take us from where we are to Who you are, that we may be warmed through and, while slowly, Surely, you thaw and make a warmer, melting love, that loves to love.

And so we come, full rounded now, not just remembering that you came and did what you have done, however much of it was unwritten and passed down, perhaps passing out of memory and dying with the People whose lives were changed, forever now, being with you in Eternity, where now you gather all who have passed into your family.

And so, Lord, let us come into your unchanging company, leaving no one, neglecting no one, overlooking no one, except those who wilfully choose to stay cold, heart unburnt, un-melted, frighteningly Unwilling to listen: the voluntarily frozen whose lives are a betrayal of being human and whose hearts are out of reach of a touching love.

Webinar: "Why the Unborn Matter"

On the 6th of December, 2025, Dr. Randal Mandock and I were co-presenters on the subject of "Why the Unborn Matter". The subject of my presentation was "The Catholic Faith's Contribution to Understanding the Moment of Human Conception"[1]. In preparation for speaking, it was necessary to study and pray: to pray and study. What follows is what arose; and, therefore, for the full context, go to the presentation itself[2].

So, when it came to speaking, my opening lines were from the work of St. John Paul II, who wrote an encyclical on the Holy Spirit in which he said that the Holy Spirit descended into the passion of Jesus Christ and drew out our Redemption:

> 'The Holy Spirit as Love and Gift comes down, in a certain sense, into the very heart of the sacrifice which is offered on the Cross' (41); and, as 'Love and Gift', 'the Holy Spirit acted in a special way in this absolute self-giving of the Son of Man, in order to transform this suffering into redemptive love' (40)[3].

---

[1] ITEST: https://faithscience.org/why-unborn-matter/.

[2] Published presentation: "Why do the Unborn Matter": https://youtube.com/watch?v=t4a_54lEJXM.

[3] St. John Paul II, *Dominum et Vivificantem* : https://www.vatican.va/content/john-paul-ii/en/encyclicals/documents/hf_jp-ii_enc_18051986_dominum-et-vivificantem.html.

Similarly, we need to beg the Holy Spirit to descend into the passion that the unborn are experiencing and draw out a good for the benefit of all mankind! At the same time, drawing on a wonderful expression that Pope Leo XIVth used in an ecumenical context, we can say with him: 'Thus, encouraged and strengthened by the grace of Jesus Christ, who is the very incarnation of hope for all, we are called and sent out to bear witness to this saving truth [that there is 'one Baptism for the forgiveness of sins ... which is the very root of all Christian fraternity'] with edifying words and charitable deeds'[4]. And, in view of all that has been said about the terrible loss or freezing of life in this generation, we need that 'hope against hope' that is of God (cf. CCC, 1819).

Finally, Pope Leo expresses a heartfelt plea for the rejection of surrogacy, which turns the child into a product and exploits the mother. Furthermore, he presses for the right to life and the flourishing of the family:

> 'In [the] light of this profound vision of life as a gift to be cherished, and of the family as its responsible guardian, we categorically reject any practice that denies or exploits the origin of life and its development. Among these is abortion, which cuts short a growing life and refuses to welcome the gift of life. In this regard, the Holy See expresses deep concern about projects aimed at financing cross-border mobility for the purpose of accessing the so-called "right to safe

---

[4] 19th of January, 2026, "*Address of Pope Leo XIV to an Ecumenical Delegation from Finland*": https://vatican.va/content/leo-xiv/en/speeches/2026/january/documents/20260119-finlandia.html

> abortion." It also considers it deplorable that public resources are allocated to suppress life, rather than being invested to support mothers and families. The primary objective must remain the protection of every unborn child and the effective and concrete support of every woman so that she is able to welcome life'[5].

And may the Church, in her wisdom, discern for the good of all whether or not she can endorse the adoption of frozen embryonic children and raise her voice, even more so, in the defence of embryonic human life. Amen.

---

[5] 9th of January, 2026, "*ADDRESS OF POPE LEO XIV TO MEMBERS OF THE DIPLOMATIC CORPS ACCREDITED TO THE HOLY SEE*": https://www.vatican.va/content/leo-xiv/en/speeches/2026/january/documents/20260109-corpo-diplomatico.html.

Where to Get Help? "Seek and You will Find" (Mt 7: 7) Help with Pregnancy or Post-Abortion Counselling

Dr. Pat Castle: '*78% of post-abortion mothers said if they had encountered ONE supportive person or encouraging message, they would have chosen life.*' And, therefore, he founded "Life Runners": they wear '"REMEMBER The Unborn" jerseys as a public witness in over 3,300 cities'[1].

Whatever country a person is in there will be, generally, organizations willing to help. Whether you google a parish or a diocesan website, life, pregnancy help, pro-life organizations, you will find they exist almost everywhere.

If you are afraid to ask for help pray, open a Gospel at random, and then ask for help. The four Gospels are often towards the end of the Bible and are: Matthew, Mark, Luke and John.

Focus on the Family:

https://www.focusonthefamily.com/uncategorized/jan-2026/?sfcid=003Uh00000NaPqfIAF

The Sisters of Life, America, go to "Where We Are"; they are in six states and Toronto, Canada: https://sistersoflife.org/

---

[1] ITEST "A Post-Roe World" Webinar publicity, now with the WCAT TV presentation on it: https://faithscience.org/post-roe/.

The Sisters of Life, Glasgow, go to the contact page and also look at the Links, further down: https://gospeloflifesisters.wordpress.com/pregnant-we-can-help-you/

SPUC: https://www.spuc.org.uk/About-Us/About-Us/Contact-Us See also "The Alma Mater Fund" for pregnant university students across the United Kingdom: https://www.spuc.org.uk/Article/385354/The-Alma-Mater-Fund-expands-across-the-UK.

Right to Life: https://righttolife.org.uk/contact-us

40 Days for Life: https://www.40daysforlife.com/vigil-search.aspx

The Good Counsel Network: https://www.goodcounselnet.co.uk/40-Days-for-Life

Life: https://lifecharity.org.uk/

Or contact your local Catholic parish church and ask for help!

Whatever your circumstances, may God bless and help you!

www.ingramcontent.com/pod-product-compliance
Lightning Source LLC
LaVergne TN
LVHW010729061126
851857LV00014B/1399